WAR GAME CAMPAIGNS

*Books by Donald F. Featherstone*

# WAR GAME CAMPAIGNS

Donald F. Featherstone

STANLEY PAUL / LONDON

STANLEY PAUL & CO LTD
3 Fitzroy Square, London W1

An imprint of the Hutchinson Publishing Group

London Melbourne Sydney Auckland
Wellington Johannesburg Cape Town
and agencies throughout the world

First published July 1970
Second impression September 1972
Third impression November 1974




Printed in Great Britain by litho by The Anchor Press Ltd
and bound by Wm Brendon & Son Ltd
both of Tiptree, Essex

ISBN 0 09 102490 0

# CONTENTS

## Contents

# ILLUSTRATIONS

# Introduction

This book is a logical step forward on the author's other war-games books published by Stanley Paul. It caters for the experienced war-gamer who is seeking larger-scaled battles and campaigns embodying greater realism in accurately representing the tactics of specific periods. And yet it is not too advanced for the relative beginner who will learn much to improve and embellish his table-top battles.

A secondary title for the book could be 'War Games For All' because it covers a large number of battles and campaigns that can be fought 'solo', by two, by three or four players, and by larger numbers. New war-games clubs are forming by the week, each with perhaps ten to fifty members. After the initial enthusiasm of battling with new opponents in different periods has palled, these clubs will need to discover new attractions and fresh ideas if they are to survive. This book contains suggestions of battles and campaigns involving large numbers of war-gamers sufficient to keep any club going for months or even years.

In this book no attempt is made to give comprehensive sets of rules for the various battles and campaigns that are described, although many of them were most satisfactorily fought to one or other of the sets of rules published by the magazine *War-gamer's Newsletter*. It is the belief of the author that any sort of standardisation of rules for war-gaming is basically unsound. The rules by which one moves model soldiers in battle on a table-top to a large degree reflect the character and temperament of their devisor—the man who is venturous and thrustful in real life will have rules that encourage attacking and the taking of chances, whilst the steady, plodding man will have rules with built-in factors probably favouring defence. The suggestions and ideas propounded in each battle account are only suggestions that might improve rules or add to enjoyment of the battle. In every case they have been actually tried and tested during the battle under discussion. Some of them were adopted as a result of the favourable impression made during the battle and embodied into our rules and now form an integral part of the system under which we fight our war games. Many others were discarded, perhaps because they were too time-consuming, fiddling, unnecessary or they did not fit our temperaments.

Every battle and campaign described in this book has actually been fought on the table-top with the author participating. Photographs were taken of many of the battles at the time they were fought and in other cases salient features of a battle have been reconstructed and photographed especially for the book. Every picture shows model soldiers from the author's collection —none of them are exclusive, the majority of them can be obtained at a reasonable cost or made or converted by the enthusiastic collector. It is with no little sense of pleasure that the author realises that he can mount a campaign in almost any period of history from the figures that gaily adorn the shelves of his war-games room.

It should be clearly understood from the start that what we are talking about here are war *games*—battles fought with model soldiers on a table-top made up with trees, rivers, roads, houses, etc., etc., to represent a battle terrain. The object of the exercise is enjoyment and relaxation coupled with mild intellectual stimulation. The alternative is an intense approach with overly serious competition and rivalry inevitably leading to arguments and bad feeling. The pundits may talk largely of '. . . re-creating the tactical ability and weapon capabilities of armies of a chosen period'. It should be realised from the very beginning that it is a game we are playing which can, for obvious reasons, bear only the most superficial resemblance to anything that takes place on a real-life battlefield.

Thanks are due to war-gamer John Bulgin, an enthusiastic London General for allowing use of his ideas and material. Another war-gamer, Bill Jenkinson, drew the maps, and Paul Reeves, the photographer, may yet succumb to the lure of the hobby.

Finally, it should be mentioned that few if any of these battles and campaigns would have been fought without the ready co-operation of table-top opponents Tony Bath, Bill Betenson, Stephen Corbett, Neville Dickinson, Bob and Stephen Douglas, Peter Gilder, John Lawlor, John Mackenzie, Ron Miles, Carl and Nigel Mottram, Chris Scott, Keith Wickham and many other 'visiting firemen'. We had some good arguments and some great battles which it is hoped they recall as fondly as does . . .

. . . DONALD FEATHERSTONE

# How to Conduct War-game Campaigns

# I

# Methods of Map-moving

It is difficult if not almost impossible to contemplate a war-games campaign without using maps upon which the moves of the opposing forces are plotted. It is not unreasonable to say that there is no time at which the war-gamer feels more like a real general than when he is studying a spread-out map conscious that he possesses the power to send a large army manœuvring down its roads, over its hills and across its rivers to completely destroy the enemy arrayed on the far side of its contours. Poor-spirited is the war-gamer who cannot find his spine tingling as he surveys a map of the Waterloo Campaign and does not wish, with hindsight, to initiate strategic and tactical moves of his own that will reverse history.

The essence of a war-games campaign lies in its continuity, in its quality of relating one battle to another so as to achieve a set objective. Of course, a series of table-top battles can be fought, each completely separated from the other except in that the results are assessed so that one side or the other can be declared the winner. But the true campaign must essentially consist of two opposing forces, perhaps divided to the choice of their commanders, being moved at regulated rates in a manner designed to cause predestined plots to smoothly reach a success-ful culmination or to be foiled by an even more successful plot.

Having made the point that maps, whether professionally printed Ordnance Survey affairs or simple diagrammatic scrawl-ings on a sheet of quarto paper, are basic essentials for a war-games campaign, let us consider the manner in which we move upon them. At first consideration it may seem a simple matter to say that an army moves down a designated road at $x$ speed for infantry and $xx$ speed for cavalry but there is a little more to it than that. What of an army's reaction to inevitable bottlenecks such as bridges, towns with narrow winding streets or a narrow defile? With Red army moving down the map from the north and Blue army moving up the map from the south, how is it decided at what stage these armies meet? And how is the fact that they are close to each other realistically concealed from both commanders? With time, patience and carefully compiled rules it is possible to simulate on *any* map in a realistic manner the movements and manœuvres of forces both large and small so that the enemy are unaware of their location and intention until a clear-cut contact is made.

About the simplest method of carrying this out is to draw a large map including all the topographical features such as rivers, roads, hills, villages and woods, etc. If your war-games table is say 8 ft. long by 5 ft. wide then make your map large enough to to be a scaled down (1 in. = 12 in.) multiple of those figures, i.e. a map 24 in. long by 15 in. wide would cover an area of nine war-games tables. The map is now covered in a pattern of inch squares—this can either be done by drawing fine lines with a ball point or mapping pen upon the actual surface of the map or else by laying upon it, for operational purposes, a transparent plastic sheet upon which these squares have been marked.

Each general has one of these maps and can move his forces over it at pre-arranged rates of movement. As an example, assuming that one move takes a day to complete, here is a set of suggested rates of movement that will work quite well on a map scaled 1 in. on the map equals 12 in. on the war-games table.

Infantry and Field Artillery—will move 2 squares on roads and 1 square across country per move.

Cavalry and Horse Artillery—will move 3 squares on roads and 2 squares across country per move.

All arms will only move at half speed over marked hilly ground.

Trains and river boats can move at 10 squares per move, taking half a day for loading and half a day for unloading.

A further touch of realism can be added by allowing land forces to move for six days and rest on the seventh day. If they move for three days and halt on the fourth day they are allowed to add a bonus distance of 50% on all the above stated moves. If they move for one day and rest for one day (this is only permitted for four days) then they are allowed to move at double speed. If a battle is fought after such a forced march then the force concerned are only permitted to move half their normal distance in table-top moves.

Although the rates of movement are known, the method of applying them has still to be decided. Of course the opposing generals can mark them upon their maps in pencil but this does not permit any degree of concealment if marked on a single map nor does it allow for any method of establishing when two forces come into contact with each other. Both of those problems can be solved in a manner that also explains how the services of a third party as an umpire may best be utilised and how the ticklish problem of concealed map-moving can be carried out by two opposing generals lacking the services of a third man.

If an umpire is available then all that is required is a sheet of transparent plastic material the same size as the map. Each general places his transparent plate over his map and marks in troop movements with a chinagraph (grease) pencil and then both hand their transparent plates to the umpire. He places them upon his master-map and continues to do so after each move until the opposing forces have contacted each other when he informs both commanders of this fact and a table-top terrain is laid out to conform to the area of the map in which the contact took place. A similar method of movement, devised by Peter Milner, is described on pages 16 and 17 of the book *Advanced War Games*. Also on page 17 is a movement method that introduces a tactical surprise feature and may be carried out by only two war-gamers.

Perhaps the most successful and best known system of map-moving in a manner than can be successfully carried out by two players lacking the services of an umpire is the Matchbox method. The first essential is to grub around until a large number of empty matchboxes has been collected. They are then glued together to a sort of chest-of-drawers so that 36 matchboxes would form a chest 6 boxes by 6 boxes and numbered from 1 to 36 from the top left working across to the bottom right. Each box in this chest now represents a grid for the campaign map which is correspondingly gridded into 36 squares and similarly numbered. The lower numbers are at the north end of the map whilst the higher numbers are at the south. Each general has a set of plastic counters with numbers or letters marked upon them, each counter representing a part of his force. For example, counter A might represent a Brigade of Guard Infantry, counter D a Squadron of Cavalry and counter F a battery of guns. He notes on a separate chart against the numbered counters exactly what force each counter represents and then, having plotted his forces on his campaign map, puts the corresponding counters into the numbered box in the chest that represents the numbered square on the map in which that formation has been disposed. Another method is to use small slips of card on which is marked the composition of the force it represents. As the forces move to another grid square on the map so the counters (or cards) are moved to the corresponding matchbox.

A certain code of ethics governs this almost ritualistic procedure—one general turns his back whilst the other slips his counters into the matchboxes and no general who is a gentleman would even think of opening a matchbox unless it is to put a counter inside it! On opening a matchbox, should a general find that it contains one of his opponent's counters then he announces that a contact has been made and the necessary steps are taken to transfer the action on to the table-top battlefield.

There are certain refinements to this method, many of which will suggest themselves to the reader. For example, if part of the campaign map includes a large open plain (e.g. grid squares number 11, 12, 13, 14 and 21, 22, 23 and 24) where visibility exceeds 1 grid square, then for each force in the plain an additional counter is put into another pre-arranged matchbox (e.g. 21) covering the central part of the plain. By this means all forces that would in actual fact be visible to each other are known (via matchbox 21) to both generals. On the other hand, where a visibility barrier such as a range of hills, divides a grid square, then the card is not put into the appropriate matchbox until the map moves have taken it to the opponent's side of the hills.

When a contact is made, each general marks it on his map. The map being scaled 1 in.=12 in. on the war-games table, it is now possible to count the squares around the contact-point that cover on an area (on the map) 8 in. by 5 in. or 8 squares by 5 squares which can be transposed to our 8 ft. by 5 ft. war-games table. Another method is to make a plastic template 8 in. by 5 in. in size and squared in 1 in. squares in the same manner as the map. A pin is thrust through the exact centre of this template and its point is then pressed into the contact-point on the map —the war-games table is the area covered by that template which is then transposed and constructed upon the table-top battlefield.

Although dealt in much greater detail at a later stage of this chapter, it is worth considering at this point some elementary method of reflecting victory or defeat in the table-top battle upon our campaign map. One simple system is for the victorious force to move forward to occupy the ground they have won by the losing force retiring back 3 squares from their original baseline (either diagonally or straight back). The winners then move forward to occupy the ground they have won but not further forward than the losers' original baseline.

So far only a relatively basic method of map-moving has been described. While it is eminently practical in that it works quite simply and produces the desired results (up to a point) it falls far short of the far greater degree of realism that can be obtained and indeed will inevitably be reached by the war-gamer as he progresses with his campaigns. There will come a time when he wishes to fight a campaign or operation that has actually taken place. The war-gamer may wish to reconstruct the D-Day landings and subsequent fighting in Normandy or to simulate the ebb and flow of the battles in Virginia during the American Civil War for example. A memorable and lengthy campaign, the Roman Rebellion was initially carried out on the Ordnance Survey map of Roman Britain, scaled 16 miles to the inch. Obviously this scale is too small to permit any really detailed information about local topographical features so that such a map has to be broken down into a series of much smaller maps, each of which covers an area which in turn can be broken down into Terrain maps composed and scaled so as to form a set of war-games tables.

Considering this method in greater detail it can be seen that the first essential is a 'Main' map which might be that of a country or extremely large operational area such as one of the American States or the area around Waterloo. The initial map movements of the campaign take place on this Main map and can be performed in a number of ways.

1. The Main map can be stretched out and pinned to a wall or a board. As both war-gamers are carrying out these initial moves on the same map, moving takes place simultaneously when both players study the map, write down their movements and then coloured pins (representing an entire force or part of that force) are moved until they enter the same area or come into actual contact.

2. Using the 'matchboxes' system, both players move on a Main map squared-off with the same number of squares as there are matchboxes. Ideally, two maps are provided so that each commander has his own.

3. A squared-off Main map is used as in 2 above but instead of employing the 'matchbox' system, each player keeps an umpire informed as to his progress.

B

4. For this method, the Main map is not marked in squares and each war-gamer marks his own tracing and then hands it over to the umpire who can chart his progress by laying it over his own map. This is the method devised by Peter Milner and described on pages 16 and 17 of the book *Advanced War Games*.

To enable forces to move realistic and reasonable distances on a map and to avoid conditions that permit ridiculous marching rates merely to enable the forces to come more quickly into contact, it is necessary that movement rates on maps should not only be geared to the scale of the map but also to reality. The following suggested rates of movement are suitable for use in campaigns during the Horse and Musket period.

*Cavalry :* Normal marching speed is 30 miles per day.

Forced marches of 40 miles per day may be carried out but 10% of the force will be liable to drop out during the march. This can be decided by a dice throw per regiment for example or by some other method to the choice of the war-gamer.

Emergency marches of 50 miles per day can be carried out when 30% of the force must be considered liable to drop out. Each such march must be followed by one day's rest.

*Infantry :* Normal marching speed is 15 miles per day.

Forced marches of 25 miles per day may be carried out but 10% of the force are liable to drop out.

Emergency marches of 30 miles per day can be attempted when 40% of the force must be considered liable to drop out.

Each emergency march must be followed by one day's rest.

*Artillery :* Normal marching speed is 20 miles per day.

Forced marches of 30 miles per day with 10% risk of dropping out and emergency marches of 40 miles per day with a 20% risk of falling out can be attempted. Each emergency march must be followed by one day's rest.

*Waggons :* As for infantry.

*Horse Artillery :* As for cavalry.

In winter deduct 10% from the daily distance and add 10% to casualty possibilities. In severe winter weather double the above amounts.

All arms will only move *half* the above distances per day when moving across country. Moving in the mountains, jungle or swamps will reduce the daily distance to one-third of the normal movement rates.

A further realistic refinement can be added by taking into

consideration the advantages of good bivouacking. Troops and horses in houses are considered to suffer no loss per night.

Troops in tents can suffer a 5% loss per night.

Troops in the open may suffer a 10% loss for each night.

In winter the above potential losses can be doubled and quadrupled during severe winter weather.

A great deal has been written about 'moving' on maps but nothing has been mentioned about the practical manner in which this is carried out. It is most uneconomical to mark movements in pencil or pen on the actual map but an ideal substitute is the marking of a transparent plastic sheet placed above the map with a chinagraph (grease) pencil or to use tracing paper and an ordinary pencil or pen. Another way of map-moving which requires no actual marking, although the positions of the troops themselves will have to be marked on a transparent sheet or a tracing, is by using a pair of dividers. This compass-like instrument can be set on a ruler first to represent infantry movement distances then cavalry, artillery or any arm whose movement-distance is to be measured. When using dividers in this manner, care must be taken that they do not 'cut' corners or carry straight on when the road curves, for example!

A device which eliminates this problem and seems to present an ideal solution to the war-gamer's problem of moving on maps is the little wheeled device that is rolled over a road on the map and which 'clicks' when each mile is traversed. These can usually be obtained in good-class stationers but might be expensive.

When moving on the Main map, time can be saved by allowing a week's move to take place at a time. Of course, when transferring to an Area or a Terrain map it will be necessary to restrict the movement to a day's march. When moving by the week on the Main map it will be possible to work out at what stage the action was transferred to the Area map. For example if a force moved half a map-move on the Main map then it would be at noon on the fourth day that they transferred to the Area map, where $3\frac{1}{2}$ days remain for manœuvring before another Main map-move becomes due.

When two opposing forces enter the same square on a Main map (or arrive in the same matchbox representing a square) then there is a possibility of a contact between those forces. To better determine the actual point at which they met and the surrounding topographical features, an 'Area' map must be made. This is an essential because the scale of the Main map is too small to allow advantages to be taken of the natural features of the countryside so that not only are both commanders working at a disadvantage but it also becomes a matter of guesswork to lay out the appropriate terrain on the war-games table. The Area map can be exactly as the map described at the commencement of this chapter when dealing with basic methods of map-moving. It should be fairly detailed as to topographical features such as hills, rivers, roads, bridges, buildings, etc., etc., and scaled 1 in. to 12 in. on the war-games table. By using this scale it is possible for the map to represent a number of war-games tables formed into a block so that a table of 8 ft. by 5 ft. for example will equal an area 8 by 5 squares on the map.

A similar map can be made but instead of being scaled at 1 in. = 12 in. on the war-games table it should be scaled so that 1 square on the map = an entire war-games table so that an Area map of 4 squares by 3 squares would represent twelve separate war-games tables.

Movement on the Area map is by the same scale as table movement which means that infantry moving 12 in. on the war-games table would move 1 in. on the Area map. This is a most important feature of the system because it is the means by which 'off the map' movements can be made. As a simple example let us picture three war-games tables adjoining each other in a line with the battle taking place in the centre table. Force A is at the extreme lefthand end of the block and it is marching to take part in the battle that is proceeding in the centre square. Marching at an infantry rate of 12 in. per move, it will take this force 8 moves to reach the lefthand end of the centre square so that it will come into the battle at the beginning of the 9th move. During this period, the battle in the centre square will have been ensuing for 9 moves. By knowing these off the map movement rates, outflanking movements are made possible with only their commander (and possibly the umpire if one is being used) aware of the actual moment at which they will arrive on an enemy's flank.

Movements on the Area map are done in the same manner as movements on the Main map—by matchboxes, tracings, clicking device, etc.

When opposing forces contact each other on an Area map squared so that, for example, 8 by 5 squares equal a table 8 ft. by 5 ft., then it is not always desirous or realistic for the battle that is to follow to be arbitrarily designated to any chance set of squares. It is better to use a transparent plastic template size 8 in. by 5 in. and squared to conform to the Area map. A pin is pushed through the centre point of the template into the exact point of contact on the Area map so that the template covers the terrain that has to be constructed on the war-games table (see ch. 18 for further details of this method). In the case of an Area map scaled so that 1 square equals an entire war-games table, the ensuing battle is fought on a war-games table-top terrain denoted by the box in which both counters repose. When tracings are used on maps that have not been squared, the point of the pin at the centre of the template is the point of contact and the battlefield that area covered by the template.

Basically, enough has been said to enable the war-gamer to fight campaigns involving the use of maps. Nevertheless only the fringe of this aspect of the hobby has been scratched because there are other methods of map-moving perhaps more to the taste and temperament of their devisors in addition to an almost limitless number of variations and additions that add polish and reality. For example, two forces approach each other on the Main map until they become close enough to each other to necessitate a transfer to an Area map. On the Area map each commander is at liberty to select his position within his own half of the map, occupying an area which can be covered by the transparent template. These positions are then revealed, either to an umpire or to each other. In the event of the template of each opponent overlapping then a third template is placed on top so that it covers an equal part of each commander's territory. The war-games table is then constructed to simulate the area covered by this third template.

As in real warfare, roads play a big part in war-games campaigns, sometimes forming a vital objective but invariably presenting a feature that must be considered and often fought for. On an Area map, roads can be shown as being first- or second-class roads, the former may be traversed normally in heavy weather but second class roads become muddy so that the rate of movement is halved. This lightly touches upon a highly important aspect of war-games campaigns—that of weather and its influence upon the course of the operations, a factor which will be considered at greater length elsewhere in this book and for which a detailed method has already been published in chapter 27 of the book *Advanced War Games*.

When two counters are discovered in the same matchbox or

two lines traced on a transparent map cover meet each other a contact has been made. However, this does not necessarily mean *close* contact as there might be a mountain, a hill or a forest in between the opposing forces. On contact, the umpire (or the opposing generals by mutual agreement) must refer to the map to see whether it is possible for each side to see the other. If they cannot do so and neither make any attempt to gain commanding ground then neither side need disclose anything at all about the troops they have in that area.

However, if scouts reach ground overlooking an enemy area, the enemy commander must give the composition of his force, saying that it is infantry and cavalry or just cavalry or some cavalry, infantry and artillery. Obviously the scouts cannot count the number of men and can only estimate their numbers —this is reflected by allowing the commander who is revealing the composition of his force to give an estimate of its strength which varies by 20% from the truth. That is to say he can increase the number of his forces by 20% or decrease it by the same amount if he desires for tactical reasons. It is easier if this estimation is given in points although the number of battalions or squadrons can be mentioned (plus or minus the permitted 20%). If both sides have scouts patrolling before them and those parties come into actual conflict then neither side need disclose information about their strength or composition until after the conflict between the scouts has been resolved.

There are variations to this method which may·be considered to add to its reality. For example from a hill rise a patrol can see anything in uninterrupted view for two or more complete squares in front of them or to the side. From level ground they can see anything in their uninterrupted view for one complete square. Or, with one clear square between both sides, each force must state its composition and strength with a 20% variation either way as required. If two clear squares is between forces then all they need to do is simply state the strength of their force with a 20% variation but no information as to its composition. If three clear squares exist between forces then they only need to state that there is a force present.

If one force is merely a scouting party or patrol then they can elect to remain and fight until their main body arrives. On the other hand they can withdraw and return to their main body with information of the approaching enemy. Obviously the correct role of a scouting party or patrol is to return with information and this they should attempt to do. On the other hand it may be that one patrol considers it to their advantage to prevent the enemy patrol returning with information and so they bring about a conflict. If all of one side's scouts are killed in the ensuing mêlée then no information need be divulged to that side which merely knows that an unknown enemy force is in the vicinity. But if any scouts escape from the mêlée and return to their own main body then they must be given information as to the size and perhaps composition of the approaching enemy force.

It has frequently been stressed that Time is the least expendable commodity that the war-gamer owns so that he may well be very resentful at having to spend a valuable evening settling a small scale battle between perhaps half a dozen men. There are ways around this—for example, the more easily satisfied amongst us may throw a dice and allow a decision to be made in accordance with its score. On the other hand it might be considered an interesting exercise to have a small-scale cavalry affair with sabres slashing, horses rearing and neighing and men falling heavily to the ground.

There are a number of ways in which such an affray can be simulated. One well tried method in the author's war-gaming memories consisted of setting up the full size war-games table with the terrain features that were likely to be used for the forthcoming battle between the two main bodies. The two opposing cavalry patrols began each on their respective sides of the table and, moving forward at normal movement rates, attempted to outmanœuvre and outflank each other so that one side could gain an advantage to destroy and prevent their enemy returning with the valuable information. There are one or two amendments and additions to the rules which make this small affair even more interesting—for example a move-distance bonus of perhaps 6 in. can be added to any definite 'charge-move'. With each side writing down their moves in advance in the manner described on page 21 of the book *Advanced War Games*, then it is necessary for the commander to actually write down his intention of charging before his men becomes eligible for the extra move-distance bonus. Another less complicated and subsequently less realistic method is for each commander to throw a dice at the beginning of each game-move—highest score has the choice of moving his little force first or second—a factor which enables many a tactical advantage to be obtained.

Yet another similar aspect is for each commander to throw a dice before moving and the commander who throws 1 or 2 is, for some reason beyond explanation, pinned down so that his own force does not move; a 3 or 4 allows his little force to move a normal move-distance whilst 5 or 6 allows them to move at double rate.

All this makes for a pleasant little diversion without the chore of moving large numbers of troops but it may not satisfactorily fill in those precious war-gaming hours and indeed may be considered a waste of whatever time it does occupy. A method that can lead to a full-scale, original and highly interesting affair is one that the author has used in the past during some memorable campaigns with Tony Bath. A cavalry patrol is considered to be a strong one of squadron strength (15 men including 2 officers). For the purposes of this exercise, this nominal scaled-down war-games strength of 15 pieces is increased to a more reasonable scale so that instead of 15 men equalling a squadron 15 men are counted as the strength of a Troop and ten Troops form a squadron. This gives a total strength of 150 cavalrymen split into ten Troops of 15 men each. It is most unlikely that any war-gamer will have 150 cavalry of exactly the same type so from this point on the ultra-realist need read no further because all that happens is that all the available French or British cavalry as the case may be are gathered together into one large squadron and then, by fighting in their strengths of 15 men each, form the whole unit. Now we have an interesting and colourful situation with 300 cavalry in two forces of 150 troopers each, massed on the two baselines by the side of the table. Both sides possess all the mobility of cavalry plus the faculty of being able to dismount and act as infantry, using their carbines. Even greater interest can be discovered from such an event if a single horse gun is allowed to accompany each force. The manœuvring that goes on to bring the gun into an advantageous position whilst two squadrons dismount to hold perhaps a vital bridge or stone wall make for a most interesting type of war game. And it is more than a war game because on its result depends the advantageous or otherwise disposition of each side's main bodies when they arrive on the battlefield. And in this connection, careful note must be taken of the number of game-moves occupied by this sprawling brawl so that they are marked up in the War Diary in a manner that allows the time taken to be weighed against the time it will require for the oncoming main body to arrive on the battlefield. It may well be that the main force of one of the cavalry patrols is two moves ahead of the other side's main body so that the remaining troopers of that side's force will have to hold off an entire Corps for two moves until their own reinforcements arrive.

To avoid time-wasting movements on Main and Area maps sometimes it is desirable to 'set up' a situation that will ensure a fairly rapid contact. One such method is for both commanders to be in possession of an Area map scaled 1 in.=12 in. on the war-games table and composed of a block of 3 by 3 tables or a block of 4 by 3 war-games tables. Let it be assumed that the force coming down from the North are the Attackers who are allowed to split into as many separate forces as desired with the sole condition that no one force must be stronger than *half* their total strength. Each of these forces will take up one of the 1 in. squares on the northernmost base line of the Area map and may all be grouped in a set of squares representing one war-games table or they may be split up as desired across all three or four tables. The Defender will arrange his forces in any manner he wishes in any or all of the three or four southernmost war-games tables. Both sides then commence map-moving in the normal way, using any of the methods already described, until a contact is made. At this point there are a number of possible variations:

1. The Attacker may refuse the contact and withdraw his forces—this will be because it is a contact not in accordance with his planned Main route of attack.
2. The Defender may refuse the contact and withdraw—because it is a line of attack against which he is ill-positioned to defend.
3. Both sides will accept the contact, in which case it has to be decided whether the battle is fought in the exact war-games table in which the contact takes place or whether a template is laid with its centre pinpoint through the point of contact so giving a war-games table made up of parts of two or more other tables.

The contact accepted by the Attacker need not necessarily be his main route of attack, he may be attempting by a feint attack to draw his opponent's forces towards that war-games table and away from positions already plotted on the map in other of the war-games tables. The Defender has to decide whether this is the case at the moment of accepting contact. He can then fight out the ensuing battle with the forces he has in that area or he may move, at recognised rates of movement, troops into the area from other parts of the Area map.

In the event of this attack being a feint, then it is fought out until it reaches a point where the Defender becomes aware of this or the Attacker acknowledges the fact. Ideally, map-moving will have been taking place in other parts of the Area map during the battle—each side moving their forces on the table during each game-move and then marking their maps up. An alternative but less realistic method is for the number of game-moves that have taken place during the actual battle to be counted at the moment the battle is frozen and then the same number of map moves are carried out. This must be done one move at a time and dispositions checked each move for a contact because it may well be that a second contact was made and an action began at some intermediate stage during the course of the action that has already been fought. Admittedly things get a little complicated at this stage but a level head backed by good-will and a deep sense of friendship usually permit the opposing generals to sort it out without resorting to warfare that is neither on the map nor the war games-table!

A similar but even more artificially set-up situation requires the Area map to have at least four roads traversing its width from north to south. The Attacker splits his force into three, none of which must be stronger than half his total strength, and secretly selects three of the four routes along which he will attack. Obviously, the Defender has to dispose his forces so as to defend all four of the possible attacking routes. This means that when a contact is made and accepted that the Defender will have to decide which force or forces he intends to move on the map so as to reinforce his out numbered defenders on the war-games table. In real war it is the object of each commander to make every effort to confront his opponent with a much stronger force, bring him to battle and decisively beat him. As will be discussed later, this does not always make a particularly interesting war game. Nevertheless, if a campaign is to achieve any degree of realism or accuracy then an occasion will inevitably occur when one commander has achieved this feat of greatly outnumbering his enemy in some specific area. Up to now accepting a contact has been optional and indeed this is the case in the method under description although an indication is given of a manner in which battle can be forced upon an un-willing opponent.

In this method no disclosure of strength or composition of a force is made until the moment when two opposing forces encounter each other on an Area map. At that point each commander informs his opponent that his force consists of either a detachment valued at 100 points or less; a Wing valued at 200 points or less or an army valued at more than 200 points (the method of assessing an army's strength by points is well known and has been explained in other publications besides the pages of this book). Briefly, an infantryman equals 1 point, a cavalryman equals 2 points and a gun equals 10 points with additions or subtractions for élite troops such as Guard and ill-trained troops such as militia or local levies. At this stage either commander can refuse battle and retire. He may move back the remainder of his map-move if the contact was made at an intermediate stage of that move or he may have to wait until the following move before he can withdraw. If desired, his opponent may follow and in this case, should he wish to force battle upon his probably weaker opponent, he can do so by cutting the enemy's line of retreat by means of a fast or forced march whilst his opponent moves at normal speed. Once an action is decided upon, either by mutual agreement or because one side cannot escape, operations are transferred from the Area map to a Terrain map and war-games table. An interesting sidelight to this method can readily be seen in that a wily commander, confessing to a weaker force, can withdraw and so draw his opponent on to a stronger force already positioned to meet them.

The importance of roads has already been mentioned—here is a method by which this can be reflected in a war-games campaign. Assuming that the opening moves of a campaign have taken place on a Main map, the action now moves on to an Area map. Now it is ruled that no force stronger than 200 points may move on the same road. This is a reasonable ruling and represents a factor that few war-gamers seem to consider. An army or even a brigade marching along a road spreads over a very wide area but in war-games campaigns most war-gamers blithely allow a huge force to go along a road all in one big lump and, when contact is made, allow their entire force to come into action as though they were all in the front rank! This is completely unreal—many books written about the British army in India during the last century tell how forces (during the Indian Mutiny for example) were spread over as much as ten miles of road. As this often represented the distance that could be marched in a day because of the heat and dust, it meant that the rear part of the column was only leaving the previous day's bivouac area when the front part had arrived at the area selected for the next night! With armies split up into groups of no more than 200 points and marching on separate roads, or even separated into advance guard; main body and rear guard marching on the same road, a new sense of reality is achieved and a far greater degree of care is required when map-moving. This can be very ably simulated by using the matchbox system—each side will deploy on to the Area map, occupying up to 5 squares deep from its base line with counters representing each force placed in the appropriate matchboxes.

The war-gamer's greatest enemy is not his opponent, it is Time. Perhaps one of the greatest attractions of this hobby lies in that it is a world of fancy strongly tinged with fact that two war-gamers enter when, free from the pressures of home and business, they get together for an evening's war-gaming. It is reasonable to say that most war-gamers prefer fighting on the table-top to moving on the map although they find fascination and pleasure in outwitting their opponent by secret tactical moves on map and in matchbox. But few will deny that they bitterly begrudge the time required to be spent when action-moves from the Main map to the Area map and topographical features have to be studied and then carefully drawn in duplicate before any further action can take place. One method of avoiding this irritating waste of time is to prepare in advance a number of Terrain maps. They can be done in pairs with a carbon paper between the sheets and several made for each type of country giving a representative set that includes plain land, hilly country, cultivated areas, desert, etc., etc. Each map must be gridded into 1 in. squares and no roads or major rivers are included. When the action moves from the Main map to the Area map, it is not difficult to decide the general type of terrain in the area at which the transference has been made. Select one of the Terrain maps of this type and cover it with a sheet of transparent plastic then draw on the plastic the roads and rivers. For each main road shown on the Main map at least three roads should appear on the Terrain map, some of them being secondary roads. By mutual agreement, it is possible to include some worthwhile objectives such as crossroads, bridges, villages, etc., that will provide a battle with an aim.

# 2

# Outflanking Operations on the Map

One of the snags of map-moving (depending upon from which facets of the hobby one derives the most pleasure) lies in the lengthy manœuvring that is often necessary before two opposing forces come into contact. Many war-gamers find this strategic contest of pitting their military brain against that of their opponent to be highly stimulating. To others it represents a tiresome delay between actually moving one's armies on a table-top terrain. However, if campaigns are to be fought in which armies move over large areas of ground then map-moving is essential and must be accepted as a part of the contest that can be made just as enjoyable as the actual war game itself.

Assuming that both armies are free to manœuvre, each will move at the specified movement rates towards his opponent until such time as a contact is made. Then the transparent template is placed over the adjoining portions of both forces (remembering that the width of the table must be between them and not the length) and the area under which the template represents the terrain to be constructed on the war-games table. Whilst both armies are having their initial moves written out or being laid on the table, those detached forces that are beyond the area covered by the template can be marching at map-movement rates towards the actual battleground so as to reach it after the action commences.

Usually this confrontation can be classified under three principal headings, dependent upon the conditions and time at which the contact was made.

1. An Encounter battle.
2. A Set-Piece battle.
3. Battle in a Prepared Position.

An Encounter battle occurs when two forces are moving relatively blindly towards each other and clash upon ground not necessarily of other's choosing. In some cases, advanced guards have met and commenced fighting until their main bodies arrive and are drawn piecemeal into the conflict—a classical battle of this type was Gettysburg.

A Set-Piece battle occurs when two armies knowingly approach each other, acting with considerable caution and circumspect until, as if by tacit agreement, they form up on a piece of ground that apparently strikes both commanders as being ideally suited for him to defeat his opponent.

Waterloo could be classified as such a contest although in many ways it overlaps into the third classification. Perhaps the first battle of Bull Run, the opening encounter of the American Civil War, falls more under the heading of a Set-Piece battle.

Sometimes circumstances decide that one army remains upon ground of its own choosing while the enemy move up to attack their position. For example, if a force is covering the only invasion route and is numerically inferior then obviously it would not move out of its selected position but would attempt to hold its chosen ground. The battles for Cassino in the Second World War are ideal examples of the defence of prepared positions.

So far as war-gaming is concerned, the first two classifications fall fairly readily into place as situations frequently encountered upon the table-top. The third classification involving prepared positions presents certain difficulties but holds out considerable promise for tactical manœuvring both on maps and table-top that provide a new dimension to our games.

Let us assume that map-moving has been going on for some time with Red force commander unaware that Blue force ceased manœuvring two moves earlier—a situation which gives them the right to be considered to be in a prepared position. It may well be that the Blue force commander has found, on the map, what he considers to be a highly promising defensive position and considers his greatest chance of success lies in staying put and awaiting attack. Such a situation can be represented in a number of ways. By remaining stationary for two map-moves, the Blue army have staked their undeniable claim to be in a prepared position which may be of such obvious quality that when contact is made and the template put into position, the merits of the position are readily emphasised on the subsequently drawn Terrain map. On the other hand, if the Area map is not particularly well defined or to a scale too small for such obvious definitions, then the Blue army commander gains an advantage from his tactical manœuvre by being permitted to draw his own

Terrain map. Of course this gives him a considerable amount of leeway but he is not permitted to rest *both* flanks on unturnable obstacles unless such obstacles actually appear on the Area map. In the event of two such natural obstacles presenting themselves, then Blue army can lay down in a prepared position and/or up to the halfway line on the table. Then the attacker deploys on his baseline. If desired and when initial dispositions are written down prior to battle rather than the troops actually being laid in full view of the opposing generals, the Blue general writes down his dispositions whilst Red general similarly writes down his baseline layout.

Should Blue general's flanks not be secured then the defender writes down his dispositions and marks them on his own Terrain map whilst the attacker can make plans for a flank approach instead of a frontal attack. To do this the attacker writes down his order of march.

Now both generals each throw the dice three times. In the event of Red army (the attacker) winning all three dice throws, the defender *must* lay down his troops on the table in his marked map positions. The attacker (Red army) may then deploy his troops within 1 ft. of the Blue army (defender) flank. Thus, at the outset of the battle it is possible for Red army to roll up the defence line of Blue army by a flank attack.

Should Red army (the attacker) win two of the dice throws, then the defender will again lay his troops down in his marked map positions but this time the attacker, whilst still deploying for a flank attack, must lay down on the flanking baseline.

If Blue army (the defender) wins two throws, then he can change front to flank to meet the attack and the battle that now takes place is fought up and down the length of the table rather than across its width.

In the event of Blue army (the defender) winning all three dice throws he is permitted to attack the enemy as they move around his flank. Red army will lay out in order of march in accordance with his written dispositions, thus presenting his own flank to Blue army's counter-attack. His flanking column must be laid down on the war-games table within 1 in. of the enemy position.

This dependence upon the luck of three dice throws may not be to the liking of the combatants. An alternative and simpler method can be used when Blue army has been stationary for a move and Red army has approached from a flank. Now Blue army has the choice of two alternatives—(a) it can change front to flank to meet the new attack—in which case it occupies a position within its own half of the board. Or, Blue army can move forward as soon as the flanking movement commences, when the terrain will be laid around the halfway point where the armies would clash and both will commence battle by deploying on their base-lines.

The moving on Area maps to stimulate flanking marches is a somewhat tricky business although highly rewarding if it comes off. Such outflanking forces must be given explicit orders as to their route and intentions, their moves must be plotted beforehand and marked on the Area map. Such moves can either be represented by the use of the matchbox system or by tracing on a transparent map cover—in either event, the services of an umpire can save a great deal of argument when the horrified defender is told that he has been outflanked! Life must not be made too easy for this outflanking force—for example, if the situation alters and the overall commander wishes to change his plans and alter their instructions he can only do this by sending a courier who will be subject to all the hazards laid down for messengers in chapter 19 of the book *Advanced War Games*. The flanking force itself will move at rates conditional upon whether it is moving upon roads, across country or by forced march. In the latter circumstance, some sort of penalty must be devised so that a fatigue reaction is reflected when they arrive on the battlefield. The flanking force can be subject to Chance Cards detailing such eventualities as slow marching, losing the way, attacks by guerillas, etc. Other factors to be considered are the effect of weather conditions upon the outflanking movement and the rating of the commander of the outflanking force—a Below-Average commander might well completely ruin any element of surprise by over-caution. In case all that is written seems to be too hard on the unfortunate flanking force then mercy can be shown by a dice being thrown before each of their map-moves so that only a score of 1 or 2 renders it necessary for them to draw a Chance Card.

Obviously all the complications outlined in this chapter are not going to be to the liking of every war-gamer. In mitigation, the author can only say that he recalls with a glow the manner in which certain war-games campaigns became a part of his life for weeks on end—when every waking action was coloured by the grandiose tactical schemes to be worked out on the map.

Cosy winter evenings come to mind with the opposing generals facing each other across the brightly lit dining-room table, carefully screened maps in front of them whilst the nest of match-boxes stood at the end of the table awaiting the repetitious movement of numbered counters in schemes that would seal the fate of Empires and the destinies of thousands of metal and plastic model soldiers.

# 3

# Lines of Communication

Generally speaking, war-gamers carry on their table-top battles with a blithe disregard for such military refinements as Lines of Communication. On occasions there is a vague mention at the start of the battle that 'Blue Army's line of retreat will be down the road that runs from the village'. But as most war games end with one army obviously on top and the opposing commander conceding the fact there is little need for such considerations. In view of the undeniable fact that a large proportion of war games end inconclusively, because of the lack of a stated objective, vital lines of communication project themselves forward as obvious targets to be held or lost and so allow for an unarguable result to the game.

However, when fighting campaigns the whole picture changes and the greatest attention has to be paid to the routes by which an army advances or retreats and is supported with food and ammunition. Even more rarely included in war games is the problem of evacuating the wounded—the enforced consideration of this aspect of warfare can add reality to a campaign.

It is one of the main principles of strategy that an army must maintain its communications if it is not to expose itself to the gravest dangers. The longer are those lines of communication, the further the thrust into enemy territory, the more they are exposed to hostile intervention. Both in real war and in war-gaming, this means that an ever increasing proportion of a force is tied down protecting those communications. On the other hand it can be reasonably claimed that a fighting force itself protects its communications by the direction of its march and by the extent covered by its frontage.

At best communications cramp the operations of an army and hamper its commander's liberty of action, fettering him and his army to its starting point by acting as a drag on mobility. Nevertheless, lines of communication must be established, maintained and protected because of the necessity of replenishing ammunition and military stores plus the obligation of sending sick and wounded to the rear.

Of course, one can always plead that favourite excuse of war-gamers—their army is 'living off the land ... it is a flying column carrying its own stores ...'. Strictly from a war-gaming point of view this highly fluid situation is most convenient and eliminates the need to worry about logistics (considered in the book *Advanced War Games*, chapter 25). Also, it makes for a carefree, glamorous sort of battling, free of the grim facts and realities of war. There comes a time when this palls, when one feels the need to delve deeper into these realities, so forcing war-games armies to 'earn their own living' instead of enjoying the luxury of unlimited stores. Like the cowboy in Western films whose pistol never needs re-loading, so our artillery and muskets seem to possess an inexhaustible supply of ammunition!

Being independent of supplies, an army cut loose from its communications enjoys great liberty of action. It is made most suitable for our war-gaming purposes because, being able to move and turn in any direction on the map, it can only be countered on the actual field of battle. Such a self-contained force gains certain advantages from its lack of communications but also it acquires many handicaps. Perhaps the principal drawbacks lie in the fact that such a force is burdened with wounded and enough stores and ammunition to last them through the expedition. This means a lot of transport which has to be protected.

Then, an army without communications in a hostile country is in a serious plight when it meets with a reverse. Not the least important function of a line of communication is that of serving as a line of retreat in case of need—otherwise a defeated force may be destroyed in fighting its way back to where it started.

From what has already been written here on the subject, the advantages and the disadvantages of lines of communication in war-games campaigns can be readily noted by the war-gamer. His problem lies in the manner in which these facts can be simulated in table-top campaigns so that decisions are required as to the respective merits of having, or not having, recognised lines of communication. Acknowledging that, except in rare and relatively hazardous circumstances, all armies possess lines of communications then it seems to be correct to state that they must exist in our campaigns. This means that the campaign must have an in-built factor that makes communications essential and rules that penalise lack of them. At the same time, the rules should not be so rigid (and unreal) as to shackle the cour-ageous or unorthodox commander who takes off into the blue with a raiding force because it is in such colourful actions that the glamour and unfettered enjoyment of war-gaming lies. But, just as in Life you get out what you put in, so must our brave leader pay the price of his ticket to freedom!

As with everything else in this book, indeed with all the words that the author has ever written on our hobby, the suggestions that follow on how to simulate these factors are merely given to stimulate and represent only basic ideas.

Lines of communication can be marked on a campaign map in yellow. Should they be cut, no forward move will be made by the front line troops until it is restored. Wherever the line is cut from this point forward to the front, after a certain number of moves (to be decided between the contestants) the forward troops will retreat to the point at which the line of communica-tion was cut.

A force is deemed to have had its lines of communication cut when an enemy force lies directly across those lines. When this occurs, the isolated force may take the following action:

(a) It may move in accordance with move rules to take action against the force threatening its lines of communication.
(b) It may move is regulated fashion until it takes a town.
(c) It may move in regulated fashion to contact with another enemy force.

All these courses of action *must* take place within *three* map- or game-moves, otherwise the unit loses 20% of its total strength for each further map- or game-move. This is repeated for another 3 game or map-moves until the line of communication is resumed or the force is wiped out. Should it not defeat the enemy in cases (a) and (c) then it may repeat courses of action (a), (b) or (c) shown above, but losing 20% of its total strength each map-move.

If the isolated force is attempting to take a town or to defeat another enemy force and so replenish its stores, then that force is in the same position as a force that has abandoned its com-munication in that it has to live off the land and/or supplies captured from the enemy. One method of reflecting these fac-tors is demonstrated in the 'Agincourt Campaign' in this book. A force that deliberately sets out, as a 'flying column' for ex-ample, with no lines of communication must be controlled by strict rules governing the length of time it can maintain itself (i.e. one man can carry four days' food for himself and his horse) whilst waggons must accompany the force to carry stores and ammunition and to transport the wounded. The conditions under which waggons cope with these problems is detailed on page 191 of the book *Advanced War Games*.

A practical method of representing lines of communication is to use yellow-headed pins on the map (three squares between each pin) running from the Stores Depot on or near the base-line of the map forward to the front-line troops. Towns have *two* pins, whilst villages are marked with *one*. If this line of communication is cut by the enemy no forward move can be made by the front-line troops until the line of communication is restored. For each move the line remains cut, one pin is removed from the point where the cut occurred and moving towards the front-line. When all pins between these points have been removed, the forward troops must at once retreat back to their nearest line of communication pin. When the line of communication is restored, one pin per move will be replaced and the force may move forward at that rate.

A raiding force can carry two such line of communication pins, one of which will carry the force a predetermined distance so that the force must retreat when both pins are used unless they have captured others during their raid which can be used to aid their movement.

By intelligent and not-too-restricting uses of lines of communication, a new, realistic and essential dimension is brought to war-games campaigns.

# 4

# Systems of Simulating Attrition

One of the most neglected features of war-gaming consists of that period immediately following the elation or depression (according to whether you are the winner or the loser) that comes at the end of a hard-fought battle. Probably because such an occasion occurs towards the end of the evening when time is running short and opposing generals reluctantly realise that the time has come to revert to their substantive ranks of husband anxiously awaited at home by a wife and family or schoolboy with homework yet to finish. Notwithstanding such undeniable if mundane claims, the situation still exists in which a force wins or a force loses a battle and then the whole affair is consigned into a sort of limbo where the battle has no bearing upon what is to follow or what has gone before except in the memories of those who took part in it.

When fighting war-games campaigns such omissions cannot be tolerated and indeed are out of the question if the campaign is to maintain that essential continuity which reflects the difference between a campaign and a series of relatively disconnected battles. Winter sports enthusiasts know that an essential and much prized part of their holiday lies in the 'après-ski' activities —the dining, the wining and the dancing that takes place in the heady and dimly lit atmosphere of Austrian or Swiss beer cellars. Similarly, the 'after-battle' activities of a war-games campaign hold a compelling interest and a tangible bearing upon the events of the campaign sufficient to make them worthy of more than a passing interest. What is to happen to the losing army? And of equal importance, what are the winners to do that will give them just and ample reward for their labours? There are many possible variations, for example the winning commander may select the next battle terrain by moving forward and occupying the ground that they have just won whilst the losing force withdraws back either straight or diagonally any distance they choose providing it is not more than $2\frac{1}{2}$ in. measuring from the base-line (on the map) that they originally occupied. Other parts of the winner's forces who have not taken part in the recent battle may move in the normal way to join their comrades or until further contacts are made. An additional alternative is to compel the winning side to remain in their new location whilst they reorganise for one day. Another method is for the winners to place the plastic template that is the same size as a war-games table on the map in any logical position adjacent to the former battle area, but not further forward than with its front edge on the position taken up by the force that has just been defeated. The area covered by the template will be the next battle area. Both sides will then do battle with their survivors from the previous battle plus 75% reinforcements (providing that does not bring either side up to more than 250 points). Assuming that it is reasonable to anticipate that the losing side have had greater casualties than the winners, then this reinforcement will give the winners of the previous battle an edge over their opponents in the second combat.

When forces have fought in a battle, an arrangement of re-strengthening them in the manner detailed above will apply until they receive their normal reinforcements (equal to the numbers lost in the battle) who may not begin to move forward from base areas until the battle is actually over. After a battle, a depleted Corps may be replaced by another whole Corps or reinforced by points to bring it up to strength from another Corps (moving at specified map movement rates). When one or more Corps become radically reduced in value it/they may be absorbed into another Corps but points strength must not exceed 250 points.

Another 'après-battle' system designed to make battles an integral part of a campaign is to draw the Terrain map on which the battle is to be fought as the centre part of a larger map so that the actual battle is fought on a centre square whilst there are about two other squares deep on either side of the battle-field. The scale of this map can be the 1 in.= 12 in. on the war-games table that has already been suggested as a reasonable scale for map-moving. When the battle in the centre square is concluded, the losing side is deemed to have lost one-third of their total strength, worked out as follows:

Infantry count 1 point; cavalry count 2 points and guns count 10 points. Thus, a side totalling 300 points to begin with must lose 100 points on defeat, made up as desired by the loser with the exception that a captured or destroyed gun may not be counted and a unit of cavalry or infantry that has been totally destroyed must also be omitted. The loser then decides what he wants to do, having two-thirds of his army facing the winner's complete force. He may stay put, or move into one of the other squares until he finds a position in which he wishes to make a stand. He may continue moving right off this map and his movement will then take place on the larger overall map of the campaign. If the loser decides to move from the battle-field, he is given one square start; if he decides to move no further than this then the next battle is planned in the usual way in this second square. If the loser decides to continue moving then he may be pursued with both sides moving at specified movement rates.

There is great temptation in war-gaming for a defeated general to, even reluctantly, say that he concedes the battle and that is the end of it. This is not good enough for a campaign where the losing general must withdraw his troops over the baseline of the war-games table. All troops failing to reach the baseline are counted as casualties but only as half casualties (i.e. two men equal the value of one man) if troops are cut off or tied down in mêlées and surrender.

Before the campaign commences, when the War Diary is being assembled, it must be decided how many moves are occupied by a day and a night. Assuming that nightfall arrives at the conclusion of the 8th game-move in a battle, a general who sees himself to be losing may use nightfall to break off the action. Then, any mêlée in progress will cease with both sides falling back 3 in. unless one side is holding defences, in which case the attacker only will retire. Either side can now move its troops for four game-moves—if any of these moves cause clashes between troops then they must be halted with a gap of 6 in. between the two sides. These four moves can be used to withdraw troops towards or across the baseline. Such night-moves can either take place simultaneously in full view of each other on the actual war-games table; they may take place one at a time with each general writing down his moves in the normal way or they may be made by means of counters in the match-boxes or tracings on transparent map covers.

If the campaign includes a system of promoting or down-grading generals in accordance with their performances then a general who decides to withdraw and manages to get a reasonable percentage of his troops safely off the field may well affect beneficially his chances of promotion or demotion.

*Above*   The British riflemen fight the French Tirailleurs for the bridge in the Peninsula War 1810. (Chapter 11 'Individualised Wargaming')
*Below*   The French Chasseurs gallantly defend the Steiring Wendel Ironworks during the battle of Spicheren. (Chapter 15 'Re-fighting an Actual Battle')

*Above*  The Vikings attack and burn an English village. (Chapter 16 'The Viking Raid'). Note the 'concealment box' in use as men emerge from the smoke of the burning palisade
*Below*  The English Army turns at bay as the French knights thunder up. (Chapter 17 'The Agincourt Campaign')

*Above*  Colonel Boles positions his men for their last stand.
(Chapter 18 'The Fight for Alton Church')
*Below*  The Royal Roussillon Regiment and their Indian allies
attack Fort George. (Chapter 19 'The French and Indian War')

French infantry supported by artillery move forward to attack
British formations. (Chapter 20 'A Corps Campaign—Napoleonic')

An army which is reduced to one-third of its strength must decide whether it intends to continue fighting. The simplest method of discovering whether or not this is the case consists in throwing a dice when 5 or 6 mean that the force continues fighting; 3 or 4 means that it retreats with units making their way off the field in good order; 1 or 2 means that all units engaged with the enemy will surrender while the rest will flee. Dice scores can be amended by adding 2 to the dice if the force is commanded by an exceptional general; 1 to the dice if he is an average general whilst 1 is deducted from the dice if he is a below average general.

An army which flees the field must be considered to be in danger of demoralisation—again the simplest way of making a decision is to throw a dice when 1, 2 or 3 mean that the force is demoralised and scatters while 4, 5 or 6 means that it rallies and regroups, perhaps taking the number of moves denoted by the throw of the dice. An army that flees the field will abandon its baggage, supply train and all encampments to the enemy.

It might well be considered that the retiring army has a very strong incentive to move quickly in getting away from the enemy whilst the victorious force has to clear up the battlefield, collect loot and see to their wounded before taking up the pursuit. To reflect this on the war-games table, a defeated army may retire a full map-move for the move after the battle whilst the victorious army may only advance half a map-move. On the other hand, if the victorious army have sufficient cavalry remaining to harass the retreating enemy then those cavalry may be moved on the map at normal move rate and in the same manner as described elsewhere in this book for pursuits and out-flanking operations.

Attrition—the progressive wearing away of man-power—is one of the most difficult factors to represent when fighting a campaign. Actually deducting men from units that are already grossly scaled-down invariably means unbalanced units causing tactical and morale-assessment difficulties. At the same time, it is both unfair and unrealistic to permit heavily defeated armies to carry on in a campaign with their unimpaired numerical strength. Perhaps the simplest answer is to divide casualty totals after a battle into three groups and then reintroduce them a third at a time during the next three battles but this can become complicated after three or four battles have been fought and an already overloaded War Diary becomes a confused affair.

Another method devised by veteran war-gamer Ed Saunders is to have a number of boxes bearing the following labels:

Base hospital.
Field hospital.
Replacement depot.
Stores depot.
Prisoner-of-war camp.

After a battle casualties are divided as follows:

One-third dead.
One-third walking-wounded.
One-third severely wounded.

The victorious side will capture all the severely wounded of their defeated enemy plus all the prisoners captured in the battle—all of these men will be put in the prisoner-of-war box. The troops counted as dead will go to the replacement depot. The walking-wounded will move to the field hospital box and the severely wounded to the base hospital. Now, all dead soldiers are immediately released from the replacement depot and move forward towards the front as replacements, moving at laid down rates. All cannon that are damaged or destroyed will be returned to the stores depot for re-issuing. The walking-wounded, the severely wounded and the prisoners-of-war move on the map at normal or reduced rates 'as desired'. This means that it may well be possible for a vigilant enemy to follow up a victory and capture any of these parties as they laboriously work their way back to their respective 'boxes' and so deprive the enemy of future man-power.

Quite justifiably, morale can be made into the most powerful factor affecting the results of battles and campaigns on the war-games table. Morale can also be utilised to represent the inevitable attrition of campaigns. Using the following method, based on a morale-system devised by war-gamer John Bulgin, a defeat or heavy losses can be made to have an effect upon future engagements in a realistic fashion.

First, it is necessary to classify all the regiments involved in the campaign. The easiest method is to declare them all as 'Untried Troops with no battle experience' with the exception of specified Guard units. On the other hand, an army that has fought before may be initially divided into two more categories, 'Veterans' and 'Tried Troops'.

*Guard Troops* are men of those units specifically designated as Guards or élite troops. They are worth 3 points each and their officers 5 points each.

*Veteran Troops* are Line Troops who have taken part in four or more battles. They are worth $2\frac{1}{2}$ points each, their officers are worth 4 points each.

*Tried Troops* are Line Troops who have been into action. They count 2 points each, their officers are worth 3 points each.

*Untried Troops* are those men who have had no battle experience. They count 1 point each, their officers are worth 2 points each.

Thus, with formations each of 2 officers and 20 men, units can attain maximum totals of:

Guards  = 70 points
Veterans = 58 points
Tried   = 47 points
*Untried* = 24 points

Taking this further, a regiment is classified as Guards if they total from 59 to 70 points; Veterans from 47 to 58 points; Tried Troops from 25 to 46 points and Untried Troops from 1 to 24 points.

After a battle, all casualties are replaced for the next battle in the following proportions:

| Type of unit taking replacements. | Proportions of losses replaced | | | |
| --- | --- | --- | --- | --- |
| | GUARDS | VETERANS | TRIED | UNTRIED |
| Guards | $\frac{1}{3}$ | $\frac{1}{3}$ | $\frac{1}{3}$ | – |
| Veterans | – | $\frac{1}{3}$ | $\frac{1}{3}$ | $\frac{1}{3}$ |
| Tried | – | $\frac{1}{3}$ | $\frac{1}{3}$ | $\frac{1}{3}$ |
| Untried | – | – | $\frac{1}{2}$ | $\frac{1}{2}$ |

Officers are replaced in the same proportions, they cannot be classified *above* the rating of the unit, i.e. in a Veteran unit having 2 replacement officers one will be a Veteran officer and the other a 'Tried' officer.

As an example:

In their first battle, a Guards unit loses 2 officers and 12 men, leaving them with 8 Guardsmen=24 points. Replacements are as follows:

| | |
|---|---|
| 1 Guards officer  | =  5 points |
| 1 Veteran officer | =  4 points |
| 4 Guardsmen       | =12 points |
| 4 Veterans        | =10 points |
| 4 Tried Troops    | =  8 points |
| **TOTAL** | 63 points |

The unit remains a Guard unit, consisting of

  1 Guards officer
  1 Veteran officer
12 Guardsmen
  4 Veterans
  4 Tried Troops

In their next battle, they lose 1 officer and 9 men, who are proportionately assessed thus:

  1 Guards officer (decided between him and Veteran officer by a dice-throw)
5 Guardsmen
2 Veterans
2 Tried Troops

Leaving:

| | |
|---|---|
| 1 Veteran officer | =  4   points |
| 7 Guardsmen       | =21   points |
| 2 Veterans        | =  5   points |
| 2 Tried soldiers  | =  4   points |
| **TOTAL** | 34   points |

Replacements:

| | |
|---|---|
| 1 Veteran officer | =  4   points |
| 3 Guardsmen       | =  9   points |
| 3 Veterans        | =  7½ points |
| 3 Tried Troops    | =  6   points |
| **TOTAL** | 50½ points |

Regiment now composed of:

|  |  |  |
|---|---|---|
| 2 | Veteran officers = | 8 points |
| 10 | Guardsmen = | 30 points |
| 5 | Veterans = | 12½ points |
| 5 | Tried Troops = | 10 points |

TOTAL 50½ points

Regiment is now a Veteran unit and not a Guard unit.

At the commencement of a campaign, all regiments, except Guards, are recruit regiments. Thus, a recruit regiment losing 1 officer and 14 men in their first battle would progress as follows:

Original classification =24 points
Losses =16 points

TOTAL 8 points

Replacements:

|  |  |  |
|---|---|---|
| 1 | Recruit officer = | 2 points |
| 7 | Tried Troops = | 14 points |
| 7 | Recruits = | 7 points |

TOTAL 31 points

Unit is now a Tried unit. (A derivation would be to count recruits surviving first battle as 'Tried Troops' so that the following would result)

Remainder after battle:

1 Officer = 3 points ⎱ giving them
6 Men = 12 points ⎰ 'Tried' ratings
Replacements as above =21 points

TAOTL 36 points

Initially, this might not appear to cause very much difference in the end-results, but as the campaign progresses so will units more rapidly rise in status as they acquire battle experience.

Various methods may be used to implement the effect in battle of these classifications. The totals derived from Morale Charts may have additions or substractions to represent the various grades. A more simple method lies in a unit having to make a morale-total of 4 or more, using specified dice.

Red dice for Guards, bearing numbers 3, 4, 5, 5, 6, 6.
Green dice for Veterans, bearing numbers 3, 3, 3, 4, 5, 6.
White dice for Tried Troops, bearing numbers 1, 2, 3, 3, 4, 5.
Black dice for Untried Troops, bearing numbers 2, 2, 3, 3, 4, 4.

# 5

## The Use of War Diaries

No war-games campaign can successfully be carried out without a considerable amount of recording. The larger and more involved the campaign the more necessary it becomes to keep a check on the whereabouts of armies or detachments, the time of arrival of flanking forces, the effect of weather, etc., etc. Scraps of paper get lost and campaigns can be held up or ruined because such a mislaid document causes some vital tactical manœuvre to become a point of dispute between the opposing generals. The ideal method of recording such events in the form of a War Diary is to use a desk diary in which each day occupies perhaps a third or half a page. The author recalls discovering with considerable pleasure that the days of the month (if memory serves him right) for the year 1963 corresponded exactly with those of the year 1863. Having a large-scale American Civil War campaign in hand at that moment, it was a simple matter to purchase two or three desk diaries at greatly reduced prices (the year had progressed through to July or August at the time) and use them for recording the events of the campaign. Otherwise, sheets of ruled foolscap are very suitable, providing a margin remains on the lefthand side in which the day is recorded whilst adequate space should remain on the rest of the paper to write down the events of that day.

Not only does a War Diary play an essential part in the working of a campaign but it also remains for posterity as a blow-by-blow account of the movements and battles that, spread over many 'campaign and actual' days until victory was finally achieved by one or other of the opposing commanders. All the campaigns and battles detailed in this book have been taken from War Diaries used at the time—the nostalgia and correction in retrospect of numerous errors has considerably added to the length of time this book has taken in its preparation!

Whilst its purpose speaks for itself, it is worthwhile detailing what must be shown in a War Diary. It must include *all* moves by an army or a detached section of that force. These need not be given in detail but can be marked by means of the names of towns, map references or the numbers of the matchboxes into which counters were placed. Make sure that all maps are carefully numbered and those numbers included in the War Diary along with the entry showing movement on that map. Just as in real life when a detached force is sometimes known by the name of its commander (i.e. Jockforce in World War II) so as to make for easy recognition.

The exact time of the year must be established before the campaign starts so that weather conditions can be carefully recorded before each day's map-move, the state of the weather must be known although this does not necessarily mean that the weather alters with every move (i.e. every day) or that it need be checked for every move. Weather conditions for all periods of the year and the effect of these varying conditions upon both map and table-top manœuvres are clearly explained in chapter 27 of the book *Advanced War Games*. So far as map-moving is concerned, the weather can be tested every day (by drawing a Weather Card) or it can be tested every second day or it can remain constant and only likely to change if a dice that is thrown shows a score that reflects the possible changeability of weather at the particular time of the year in which the campaign is being fought. For example, it would seem that weather is more likely to change during winter than in summer so that the dice score should render this fact more likely.

Once a contact is made and a battlefield terrain is laid, both commanders can mark their full dispositions on their Terrain maps and only after that is a dice thrown to see if there is any change in the weather. This may have a most dramatic effect on the battle in that sudden torrential rain can make bow strings wet and unfit to be used or the musket will not fire or artillery can bog down and must keep to the roads, etc., etc.

The War Diary has the greatest value in keeping check when map-moving changes from a Main map (of a country or large expanse of countryside) onto the larger scale of an Area map. It may be that different forces are moving along different routes or attempting outflanking movements on both the Main map and the Area map or else an outflanking movement is only being attempted to actually bring troops on to a table-top terrain at some intermediate stage in that battle. These are tricky things to keep check on and only by clear notes in a War Diary can this be done.

Then there is the question of commanders keeping in touch

with detached parts of their force, altering orders and generally keeping control of their armies. It may be that the war-gamer is not fussy and allows his commander a God-like ability to keep in touch with far distant units whose actual location is unknown to him. But this will not do for most war-gamers who require the same degree of reality as in real life. This necessitates the commander, perhaps unaware of the exact location of his forces, sending out messengers and couriers to them. The progress of these couriers (controlled by the rules given in chapter 19 of *Advanced War Games*) must be recorded in the War Diary so that their exact time of arrival is known and the unit to whom they are bringing orders cannot take steps in the desired direction until such orders are actually received.

If the campaign is a Club Project then each commander of a separate force will keep his own War Diary whilst the commander-in-chief keeps a War Diary of his own in which he co-ordinates all his detached forces. If possible, it is a good idea to place all commanders on their honour not to communicate their whereabouts or any facts concerning their force with their commander-in-chief so that orders have to reach them and be obeyed or disobeyed as in real life.

After a contact has been made, fighting will take place on the war-games table. This is not all that occurs because it may well be that units not yet on the table will continue to move towards the direction of the battle or to cut a line of retreat, etc. This means that they will continue to be moved in matchboxes or by tracings on the transparent map cover. If an umpire is available to co-ordinate all such movements then life is made much easier but if two war-gamers are attempting to fight a campaign with all its many aspects of secrecy and without the services of a third party then they will find a War Diary indispensable. For example, should another contact, separate from the battle already taking place, be made during moves on the map, then it is noted in the War Diary and fought in due course. Its effect in relation to timing on the battle in progress has to be considered. It might be that the map contact was made at the end of the fourth game-move in the table-top battle already in progress. Of course that battle continues and may take a further four moves before completion. Theoretically, the second battle has then been going on for four game-moves so that when the second battle finally gets on to the table-top and is fought out it may be discovered that battle number 2 actually finished at the

D

end of its third game-move (meaning that it was completed one game-move before the end of battle number 1. Should those two battles have any bearing on each other or be in the same area then this has to be worked out before the campaign can be resumed. Again a War Diary plays a vital part in such calculations.

Before commencing map moving it is a good idea to determine the direction of the wind and mark it in the War Diary. This enables a commander to follow that well-known military practice of 'marching to the sound of the guns'. In other words, if the wind is blowing from east to west and a battle is taking place in an eastern area of the map, then forces moving a specified distance away from the battle and on its westward side may be considered to have heard the sound of artillery and their commander is permitted (without orders by courier or any other notification from his C.-in-C.) to march towards the sound of the battle. This means that a very timely reinforcement may well arrive on the enemy's flank at a crucial stage of the conflict. This question of wind direction is subject to general weather conditions applying at the time and may be non-applicable during a blizzard, for example.

6

# Small Forces Holding Up Larger Forces

In real-life warfare it is the aim of every commander to manœuvre his forces in such a manner that success is assured when
they come up against the enemy because they are in overwhelming strength. This may be all right in real life but when war-
gaming it is a step which usually ruins a campaign as the resulting
battle rarely holds any interest for either general because the
smaller side has no chance whatsoever. Such a situation can be
handled by the larger force being restricted to a superiority of
not more than 25%. The surplus troops may be retained in the
rear as a reserve force or, if the surrounding countryside allows,
they can be sent around the enemy's flanks and menace his line
of retreat. This may be done in the following manner.

Red army (the larger force) has won and retained its extra
troops as a reserve, leaving them free to press the pursuit. Blue
army (the smaller force) has retired two moves on the map by a
force-march in order to escape. If Red army's reserves have
been sent around the enemy's flank, then Blue army can retire
by an alternative escape route if one exists but if all routes are
blocked then Blue army must either disperse or surrender. In
the event of Blue army winning the original battle then it may
either retire in which case Red army must halt for the next two
map-moves or Blue army can elect to fight a second action in
which Red army uses its reserves and enough of the original
force to bring it up to Blue army's strength.

If Red army has sent out a flanking force but Blue army wins
the original battle, then Blue army can turn upon the flanking
force and attack it. If this encounter is indecisive then Blue
army can either withdraw by an alternative route or if none is
available, they can attempt to fight their way through the flanking force with the main enemy army coming down on their rear
after four moves.

During the original battle Blue army have the alternative of
breaking off the action if things are going badly and they suspect
their retreat is threatened. To do this, Blue army must get one-
third of their original force back to their baseline. Having done

this, if Red army's flanking force has not yet reached its blocking position then Blue army is free to withdraw. Both armies can them move in the usual way on the map and continue the campaign.

Under certain circumstances small forces can hold up or hamper larger forces under a system governed by:

(a)  The rating of the commanders of both forces.
(b)  The strength of the two forces.

Values are given to represent the rating of a commander:

Exceptional commanders=3.
Average commanders=2.
Below average commanders=1.

Similarly, values are given at the rate of 1 per 30 points of strength of a force so that a force of 120 points=4 and a force of 60 points=2.

When a situation arises where a smaller force is attempting to delay or slow up a larger force, then the values of those two forces are totalled up and action is taken in accordance with the table below. Of course, it must be appreciated that these reactions are represented in moves on the map and not on the actual war-games table.

*In mountainous, heavily wooded or broken country* a difference of 3 between two forces means the larger force is slowed down to moving at *half* speed.

*Example.* A force of 120 points with an Exceptional commander (value 7) is opposed by a force of 30 points also with an Exceptional commander (value 4) or a force of 120 points with an Average commander value 6) is opposed by a force of 30 points with an Average commander (value 3).

A difference of 2 means that the larger force is slowed down to *quarter* speed.

*Example.* A force of 90 points with an Average commander value 5) is opposed by a force of 30 points also with an Average commander (value 3).

Where there is a difference of 1 then the larger force remain *stationary for one day*.

*Example.* Where a larger force of 120 points led by a Below Average commander (value 5) is opposed by a force of 30 points with an Exceptional commander (value 4).

Where the value of both sides is the same then the larger side remain *stationary for two days*.

*Example.* Where a force of 60 points led by a Below Average commander (value 3) is opposed by a force of 30 points with an Average commander (value 3).

When there is a difference of −1 point in favour of the smaller side, then the larger force must remain *stationary for three days.*

*Example.* A force of 60 points with a Below Average commander (value 3) is opposed by a force of 30 points with an Exceptional commander (value 4).

*In open country* where there is a difference of 2 between the forces then the larger force are reduced to moving at *half* speed.

*Example.* Where a force of 120 points led by an Average commander (value 6) is opposed by a force of 30 points led by an Exceptional commander (value 4).

Where there is a difference of 1 then the larger force are reduced to moving at *quarter* speed.

*Example.* Where a force of 90 points led by an Average commander (value 5) is opposed by a force of 30 points led by an Exceptional commander (value 4).

Where the values of both sides are the same, then the larger force will remain *stationary for one day.*

*Example.* Where a force of 60 points led by a Below Average commander (value 3) is opposed by a force of 30 points led by an Average commander (value 3).

Where there is a difference of −1 in favour of the smaller force then the larger force will remain *stationary for two days.*

*Example.* Where a force of 60 points led by a Below Average commander (value 3) is opposed by a force of 30 points led by an Exceptional commander (value 4).

This method may be applied to forces that are not exact multiples of 30 points—thus a force of 20 points could equal 1 and hold up a force of 80 points equalling 4 and so on.

The method of rating commanders used in this chapter is given in fuller detail in chapter 18 of the book *Advanced War Games* under the heading of 'Commanders Good or Bad'.

# 7

## The War-Game League

Fighting a single disconnected war game with nothing on the result can considerably detract from its interest. At the same time it is not always possible to embody that game into a campaign or to take the time or trouble to construct a larger-scale affair. Here is a method of war-gaming which will at the same time fill in a gap between campaigns and yet allow each game to have an important bearing on other games that are subsequently fought.

First a selection of six different types of terrain is devised and set out as maps scaled to the war-games table. The following will serve as an example.

1. Agricultural country with a village.
2. A town that is spread out over the entire war games table.
3. Open ground with roads, trees and cultivated enclosures.
4. A river surrounded by fields and with some marshland and bridges.
5. A hilly country with ravines and defiles.
6. A coastal terrain with beaches and inlets.

Next is devised a series of different types of battles such as:

1. An encounter battle in which two advancing forces clash with each other on ground not necessarily of their own choosing.
2. A set-piece battle in which the map is drawn to cover an area of perhaps four war-games tables set in a block. The two opposing generals position themselves deliberately so as to take advantage of ground and battle does not commence until both sides have more or less convinced themselves that the position they have chosen is the one most likely to lead to success.
3. A defensive battle in which the attackers will be superior in numbers to the defenders by one-third. To counter this the defenders may set up their own terrain. The attackers may detach a body to move around the flanks or to the rear by using map-moves (see other sections of this book dealing with such contingencies).

4. A delaying action, with Red force having twice as many as Blue force at the start but Blue force having sufficient reinforcements arriving at various times during the battle to give them an overall total of 25% superiority to Red Force.

Battles number 3 and 4 have to be fought twice with each general taking it in turns to be the defender or the attacker.

Six games will be fought to form the series and 2 points will be given for a win and 1 point for a draw plus points deducted for losses of men and guns. After say three battles a specimen League Table might read as follows:

|  | Played | Won | Drawn | Lost | Points gained* | Points lost | League points |
|---|---|---|---|---|---|---|---|
| General 'B' | 3 | 1 | 1 | 1 | 500 | 490 | 3 |
| General 'A' | 3 | 1 | 1 | 1 | 490 | 500 | 3 |

*Infantry=1 point, Cavalry=2 points, Guns=10 points, etc.

At the commencement of this 'war-games league' it will have to be decided whether the battles are to be fought in some set order or whether the six varieties shall be written on slips of paper and drawn from a hat. Likewise the same method can take place with the terrain so that a man might draw a defensive battle slip of paper but find that he has picked a piece of terrain that is not particularly suitable for that type of conflict.

This method of war-gaming has an added attraction in that the battles need not be of the same period. For example, the first battle could be a Napoleonic affair, while the second battle could be between two ancient armies and the third could involve tanks and machine-guns.

To demonstrate the operation of this type of war-gaming let it be assumed that two war-gamers have already fought three of their six battles. The first battle was a set-piece encounter in which a Bavarian force defeated a Saxon army in an 1870 period battle. General 'A' commanded Bavaria while General 'B' commanded Saxony and the Bavarians lost 110 points and the Saxons lost 140. Battle number 2 was a defensive battle in which a Roman force led by General 'A' attempted to hold off Carthaginian army (complete with elephants) commanded by General 'B'. The Romans were defeated in this encounter and almost wiped out, losing 265 points, the Carthaginians 190 points.

The third game was a Napoleonic encounter battle in which General 'A', commanding a British force, fought a hard and drawn game with General 'B's' Frenchmen. The British lost 125 points and the French lost 160 points. Thus these three battles will give us a League Table as shown on p. 45.

With a situation as tight as this, each one of the remaining three battles was going to be a very tense affair!

# War-game Clubs

# 8

# Forming a Club

Mankind is endowed with a peculiar trait which makes him desire to congregate together with his fellows. The English-speaking races translate this natural gregariousness, this herd instinct into reality by getting together to form clubs of people with similar interests be they stamp collecting or even mass drinking! Perhaps defensively it can be claimed that the war-gamer establishes and joins clubs together with his fellow hobbyist for a more tangible and constructive purpose. In the first place, war-gaming is a hobby that obviously is more suited to the joint efforts of two or more people rather than a single individual ploughing a lonely furrow. This is because it is competitive and because much of what occurs on the war-games table and the campaigns map is essentially a battle of wits in which secrecy plays no small part. For these reasons alone, the war-gamer is encouraged to seek out at least one other person of similar interests.

On top of that, there is no doubt whatsoever that in many cases a combination of new ideas and resources of a number of people add greatly to the enjoyment and true depth of the hobby. With the ever increasing range of readily available Airfix war-games figures the massing of armies is no longer an expensive business but it still takes an immeasurably long time to paint up sufficient figures to form two large enough forces to give a realistic battle or campaign. By joining with his fellows, each collecting and painting to a set plan, the war-gamer is soon able to take part in large-scale battles involving far more model soldiers than he could paint in a year.

There are other benefits such as those gained by the man or boy who does not have the necessary space to lay out war-games tables of any great size or indeed to lay them out at all. By joining a Club such a person is able to take advantage of table-top terrains set out in a variety of periods and to a standard that may well surpass anything he could himself achieve. Then there is the matter of variety—most war-gamers have a favourite period in which they collect their armies. Often it is the Napoleonic period when they attempt to build up armies of British and French, taking all their available cash and time in so doing. That same man, whilst his enthusiasm for the Napoleonic era will never wane, may also have secret or not so secret yearnings to fight battles with chariots, war elephants and Roman legions or else he may desire to try his hand at modern warfare with tanks and the other multi-weaponed forces of World War II. By joining a War-games Club he is able to indulge himself in these fancies and in return his figures can be used on Club nights by a modernist or an Ancient-fan who has always wanted to be a Wellington or a Napoleon.

The well-organised War-games Club meets at least one evening a month and probably more often. The first arrivals set up 6 ft. by 2 ft. trestle tables and begin to lay out terrain they have brought with them or else taken from the Club collection donated by members and left on the premises. When the later arrivals turn up they have the choice of fighting in perhaps any one of five or six periods so that in the course of a year a young war-gamer can sample the delights of battling in many areas before finally selecting that which appeals most to him.

Some Clubs are more fortunate than others and instead of hiring the village hall they are able to secure some permanent premises in which they can safely store their terrain and models, leave large-scale battles in situ from week to week and have far more frequent meetings than are possible when hiring fees have to be paid. It might be a cellar beneath a member's house which has to be cleaned out and turned into a Club room or an old Nissen hut set on a piece of waste land—although in this age of the Permissive Society it is feared that intruders might well viciously destroy such items as model soldiers and terrain which are of no value to them but of inestimable value to their owners. A permanent Club room also allows the Club to have a small library or an information service, with members providing spare books from their own collection. Joint subscriptions can be taken to magazines of interest to war-gamers and military collectors which can be shared around the Club on a rota basis.

Perhaps the greatest boon of a War-games Club lies in the many projects made possible by its larger numbers of members and their joint facilities. Usually, as already mentioned, a war-games evening finds a number of tables in varying periods set up and awaiting combatants. Inevitably there comes the week when holidays intervene or human nature takes over and everyone leaves it to everyone else to bring something along so that games have to be played with very small numbers or hasty journeys home are made to bring back enough troops to set up a table. This can be avoided by having a Club Project that runs throughout the year in addition to the other activities of the Club. Such a project should be carefully chosen after consultation in which all members of the Club have a voice; it should be made mutually interesting and possess a built-in factor that allows its competitive interests to run from month to month over a long period without any diminishing of interest. Perhaps the simplest way of tackling such a situation is to have a large map made up of perhaps a dozen war-games tables. Two commanders-in-chief are selected, who deploy their forces on the map and then co-ordinate the results of each battle and carry on accordingly from month to month. The method of fighting is for one of the war-games tables to be set up at each meeting of the Club and for a specified general (each member of the Club takes a turn at this) to command a force under the direct orders of his C.-in-C. At the end of the battle the result, together with losses, etc., is handed to each commander-in-chief and to the umpire who marks them up on his map. At the next meeting another one of the war-games tables is constructed and a second general fights a battle in that area. He gives his results and losses, etc., to his C.-in-C. and so the procedure goes on month by month. The simple truth about such a project is that the commanders-in-chief have perhaps 10 regiments at their disposal in all but as those 10 regiments are used in one sector in the first week and the same 10 in another sector in the second meeting and so on it is possible for 10 regiments to be multiplied by say 12 monthly battles so that each commander-in-chief commands (on paper) a force of 120 regiments which are deployed throughout his map. This is a simple method of 'audience-participation'—a factor which should be the aim of every War-games Club because it is the kiss of death to have young members or shy members coming along to meetings and standing by without taking part or being welcomed into any game.

The well-known American war-gamer Fred Vietmeyer regularly presides over very large-scale Napoleonic games, usually reconstructions of actual battles of the Napoleonic Wars. These games involve as many as 20 or 30 war-gamers, each of whom has a command and who receives his battle orders and then fights accordingly. These commands are graded according to the experience of the war-gamer who starts at the bottom with a small command and progresses slowly up the ladder of promotion until he commands a Division or Corps. An excellent account of one such battle, Vittoria 1813, from its very beginning right through to its award of battle honours and promotions after the smoke and dust had died down, appeared in *War-gamer's Newsletter* for May, June and July 1969.

Perhaps it is not entirely to the taste of the conservative English war-gamer to actually bear a military rank in the Club and to seriously accept promotion or demotion according to his prowess on the war-games table. Nevertheless every man has a competitive bone in his body so that the stimulus and interest of rising or falling in some published League Table has an appeal. If Club Projects involve certain designated regiments or formations provided by individual war-gamers, so that Charlie Jones is well known as the brave commander of the 37th Regiment of Foot, the 88th Regiment of Foot and the Scots Greys, then any battle honours gained by those regiments will add lustre to the name of Charlie Jones. The manner in which this can be demonstrated is for the Club artist to paint a series of small outlines of flags on a large sheet of cartridge paper. Each regiment will have its own flag on which will be marked those battles at which it distinguishes itself during Club war games.

To get a War-games Club off the ground is not an easy matter—to arouse a 'Club spirit' or sense of belonging to a group of individuals of varying ages and occupations requires a focal aim to serve as a banner on which all can be rallied. A novel method with a merit of being inexpensive can be the formation of a Club war-games army. First every member of the Club is asked to donate the sum of 2*s*. 9*d*.—the price of a box of Airfix figures—and as an example let us assume that sixteen 2*s*. 9*d*.'s have been collected. Hold a meeting at which the project is explained, when everyone is told that the Club are going to buy 16 boxes of Airfix figures, and that each member will be responsible for painting up one box of figures which will then become the Club property and will always be available to be used on war-games evenings. If there are no strong objections perhaps the easiest period to take is the American Civil War for which four boxes of Federal Infantry, two boxes of Cavalry and two boxes of Artillery are purchased. The same numbers are purchased for the Confederate forces. Everyone knows that these boxes of Airfix figures contain about 40 figures or so in a large variety of positions, by taking all the men firing and place them together and all the men who are running, and all the men kneeling, etc., etc., one can obtain regiments of say 20 men all in firing positions or 40 men all in kneeling positions together with officers and standard bearers. Divide the forces up in this manner and allocate one regiment to each man who is then responsible for painting it and bringing it along to the Club. On the duly designated evening, if everyone has played up to form, two small armies of Federals and Confederates will be ready for action. In addition to the fact that everyone is made to feel that they are a part of the project, there is also the interest in comparing styles of painting and perhaps, in battles that follow, proudly handling or following the fortunes of a regiment painted by oneself.

Now follows, at some length, a series of reasonably detailed war-games campaigns and similar activities suitable for taking up as Club Projects. The Wessex War-games Club of Southampton were the first war-gamers seriously to take up the English Civil War (1642 to 1651). A colourful period that has much to offer the war-gamer, for some inexplicable reason it has not been explored in a war-gaming sense. Stimulation was given to the period as a project by the combined efforts of the author and Major Alastair Malpas who was a Club enthusiast whilst stationed in Winchester on a military course. After touring many English Civil War battlefields in the south of England, the pair together devised the campaign that is described here and the rules under which it was fought. Major Malpas acted as campaign co-ordinator throughout, although leaving the area and having to carry on his duties by mail.

At a meeting called to discuss the project it was agreed that every member of the Club who wished to participate should declare for King or Parliament. This pledged him to assemble a small personal force that would participate in regular war games at the monthly meetings of the Club. These forces were as follows:

*A Foot unit* to consist of 18 musketeers and 18 pikemen with 2 dismounted officers and a mounted officer.

*A Cavalry unit* to consist of 12 troopers and 2 mounted officers.

*A gun crew* to consist of 5 men plus their gun. All guns were

Falcons of the same calibre and no other guns would be used except by mutual agreement or pre-arrangement.

No commander was to be allowed to have a force larger than 2 Foot regiments and 1 gun *or* 2 Cavalry regiments and 1 gun *or* 1 Foot regiment, 1 Cavalry regiment and 1 gun. Anyone coming to a meeting with forces larger than that was expected to loan them to anyone lacking troops for the duration of that battle.

The campaign was conducted by dividing England into three areas—North, Midlands and South, with the following respective start-points—Durham, Leicester and Winchester.

The Northern commanders were Prince Rupert for the Royalists and Cromwell for Parliament.

The Midland commanders were Charles I for the Royalists and Fairfax for Parliament.

The Southern commanders were Hopton for the Royalists and Waller for Parliament.

Battles were to be fought one at a time, between the forces in the respective areas. The campaign co-ordinator (Major Malpas) organised each operation, laying down the specific objective for the Captain-General to obtain. The first battle was to take place in the North, to be joined in the area of Durham. At the campaign co-ordinator's discretion, the losing side was to retreat north or south but not below Sheffield.

The total forces of the Club first represented the two forces in the North, then for the second battle they became those in the Midlands (when they were the armies of Charles and Fairfax) and for the third and southern battle they were the armies of Hopton and Waller. This meant that although six armies were in the field and their results were inter-connected, in reality it was only necessary for the Club to assemble two armies—one for King and one for Parliament.

Points were scored as a result of each battle and eventually a League Table was made out showing the respective positions of the six armies. Scoring was as follows:

For achieving objective—10 points.

A drawn battle—5 points.

For each standard captured—3 points.

For each Captain-General killed or captured—5 points.

For each junior commander killed or captured—3 points.

For each gun captured—2 points.

For causing enemy 25% casualties—2 points.

For causing enemy 50% casualties—5 points.
For causing enemy 75% casualties—7 points.
The sequence of the campaign was maintained by the two Northern armies fighting again after the Midlands and the South had fought. This battle would take place in the new area to which the losers in the North last time had retreated. Then the same procedure took place with the Midland and Southern armies. If an army continually lost and was finally forced out of its area into one of the other two areas then that army was considered destroyed and took no further part in the campaign.

The way in which the campaign was conducted on the evenings when we met to fight held several novel points. In the first place the Supreme Commander of each army, i.e. the Captain-General, was selected on the night of the battle when a draw was made between all those taking part on each side. The man so drawn as commander could not command any actual force in the forthcoming battle and had to hand his regiment or regiments over to another. He commanded through junior commanders whom he selected from those on his side and he dealt only with them and not with unit commanders. At the conclusion of each battle, the captain-general signed a pro-forma showing how the battle developed. This was sent to the campaign co-ordinator who kept records and scores.

In order that battles could be fought to a finish in an evening, no battle was to last longer than eight game-moves. At that point, night was deemed to have fallen and an assessment was made as to whom the situation favoured. An interesting and realistic situation was built into the campaign in that on the night of a battle the Royalist forces were represented only by those war-gamers who turned up with troops of that army—similarly the armies of Parliament were represented by war-gamers turning up with their Roundheads so that, just as in a rather muddled Civil War like the one we were attempting to simulate, battles took place between whoever happened to get to the field at the right time. Needless to say there was a considerable amount of lobbying and pressure put upon missing members to turn up and swell the numbers. To allow for the fact that one side might be larger than the other, the battle instructions sent down by Major Malpas were flexible in that he detailed a situation in which a smaller force defended against a larger force. On the night, if more had turned up for King than

E

Parliament, then the Cavaliers were the larger force and did the attacking whilst Parliament with the smaller force did the defending.

The routine that came to be accepted on war-games nights was for everyone to assemble at the appointed time when the respective supporters of both sides declared themselves and took their post on either side of the war-games table to cry out challenges and abusive remarks across the table. Next the Club Secretary opened the sealed instructions sent down by the campaign co-ordinator and they were read aloud. Until this point no one had any knowledge of the instructions. Then the terrain was set up in accordance with details contained in the instructions and the respective captain-generals were drawn for. They selected their junior commanders and proceeded to work out their tactics. Then the forces were laid out and the battle commenced. Eight moves later the result was decided, both commanders completed and signed the pro-forma and then everyone retired round the corner to a handy pub to re-fight the battle.

This turned out to be everything that was hoped for with at least twenty members purchasing and painting up armies for one side or the other. On the night of the battle there were never less than fourteen or sixteen players, each of whom handled his own small group on the battlefield and felt justifiably proud or ashamed of their progress.

Briefly, the campaign started off with a bang when on the very first night the supporters of Parliament turned up in force so that they outnumbered the Royalists by nearly 2–1. The campaign co-ordinator's instructions gave the smaller force a line of hedges and a house to hold until nightfall. In the battle that raged most realistically for the best part of $3\frac{1}{2}$ hours, Prince Rupert's men fought like fiends and although they suffered great casualties, by nightfall they still held the line of hedges and the house. Thus the Royalists commenced with a very fine win with bonus of 10 points.

And so, for month after month the English Civil War was re-fought in Southampton until with the campaign having gone twice round (two North, two Midland and two Southern battles fought), the situation was finally resolved. If it began with a flourish, it ended in the same way because Waller in the South had to decisively win a battle in the Winchester area to head the League and take the crown away from Prince Rupert's Cavaliers who had been top of the League since the very beginning of the campaign. Waller had had one amazing and decisive success in the first battle near Winchester when Hopton the Cavalier commander had been forced to flee the field only to run back straight into the arms of a squadron of Parliamentary cavalry who were returning to the field after having run out of control. Unfortunately, Waller did not rise to the occasion in his last battle and a rather dull and indecisive drawn affair resulted so that the Royalists actually won the English Civil War on the table-top if not in real life.

This was a very fine Club Project which aroused great enthusiasm and possessed the merit of making people take sides and support their factions in a rather serious and strong-minded manner. At times it almost became like a Civil War between the Conservative Party and the Labour Party! Although the campaign has now closed, the period of the English Civil War has strongly caught the fancy of the members of the Wessex War-games Club and there is never a war-games night at which the Cavaliers and Roundheads are not going hard at it again at some part of the room.

The following three books were studied by members of the Wessex Military Society and they are recommended as being excellent material from which to obtain a background to any war-gaming in the English Civil War period:

*Battles and Generals of the Civil Wars 1642–1651* by Colonel H. C. B. Rogers, O.B.E.

*Battles of the English Civil War* by Austin Woolrych.

*Edgehill 1642—The Campaign and the Battle* by Brigadier Peter Young, D.S.O., M.C., M.A., F.S.A., F.R.HIST.S.

# 9

## Club Activities

It has already been stressed that the one essential feature of a War-games Club must be that activities are so arranged that everyone is included. Not only should a Club be a meeting place for a group of war-gaming enthusiasts to gather together and fight battles on a larger scale than normally, it should also provide facilities for instruction preferably by demonstration. That is to say experienced war-gamers will, at specially arranged sessions, demonstrate the various known methods of firing artillery in the Napoleonic period for example or how to represent a Macedonian phalanx on a table-top battlefield. Similarly there must be scope for the lighter, less serious side of the hobby because, contrary to the solemn edicts pronounced by intense and often youthful enthusiasts, there is a place in the hobby for sheer enjoyment.

Conscious of this fact, the Wessex Military Society have devised a method whereby smaller and lesser publicised facets of war-gaming receive an airing. At their monthly Saturday afternoon meeting, the proceedings commence with a talk on some suitable military subject that lasts for about an hour. Tea and biscuits follow and then for the last hour and a half of the meeting some form of demonstration usually connected with war games is staged. Devised by members of the Club, these demonstrations are invariably practical, informative and amusing. For example, one memorable spectacle, entitled 'Instant Guerillas', was a Viet-Nam type of affair when numbers of ragged Viet-Cong (cleverly converted from Airfix figures) did their dirty work and then vanishing rapidly into native houses, emerged on the other side as civilians! Then there was a demonstration of siege warfare during the 17th century, involving a highly ingenious fort made from polystyrene tiles which jig-sawed together and could be taken apart to allow for breaches made by cannon-balls. Modern warfare enthusiast Ron Miles showed just how successful could be the paratroop drops detailed in chapter 6 of the book *Air War Games*.*

* *Air War Games* by Donald Featherstone, Stanley Paul, 1966.

Believing in the benefits of 'audience participation', the author devised a short game for one of these demonstration periods. Called 'Individualised War-gaming' it involved every person present and not only caused amusement but appeared to have potentialities for actual war games when a similar situation had to be solved. Basically, the idea consisted of two small forces—one British and one French each consisting of a lieutenant, a sergeant, a bugler and ten riflemen. One of the riflemen was a 'strong man'! in other words, he was one of those exceptional men one occasionally finds in platoons who are not only bigger and stronger than their comrades but they are also better shots, bayonet fighters, etc., etc. Each figure had a small number painted in white on its base and at the commencement of the game all persons present were divided into two equal parties, a responsible and experienced war-gamer was nominated as the lieutenant and given the figure of the lieutenant. The lieutenant nominated his sergeant and his bugler to whom numbered figures were given and then each of the riflemen was given a figure. This meant that every person in the room was actually represented individually on the war-games table. A terrain was laid on a trestle table 6 ft. long by 2 ft. wide—it consisted of a river running right down the middle of the table from one end to the other lengthways which ended in a bridge at the furthest end. On either side of the river were small clumps of trees, mounds of earth, fallen trees etc., to give cover to the riflemen.

As this game proceeded very smoothly everyone appeared to readily understand the typed instruction sheet they were given (together with a 12 in. strip of stiff card marked off in inches like a ruler, plus a dice). It is considered adequate to give an exact copy herewith of that instruction sheet.


## INDIVIDUALISED WAR-GAMING


*Narrative*

It is in the Peninsula in 1810. A small detachment of British Riflemen, formed of a lieutenant, a sergeant, a bugler and riflemen are making their way back from the French side of a river, over which there is only one bridge. By a strange coincidence, a force of French Tirailleurs of the same strength have been trapped on the British side of the river and are seeking to cross and gain their own lines. Both forces will need to cross by the same bridge.

Each man is worth 1 point.

The Lieutenant and the 'Strong man' are worth 2 points.

## Movement

9 in. normal.

12 in. running (only alternate moves).

If a man is wounded, he deducts 3 in. from his move-distance, i.e. 6 in.

When a wounded man is aided by a comrade, both together they move 9 in. but neither can fire whilst moving.

'Strong man' always moves 12 in.

## Firing

Range 12 in.

It takes a man $4\frac{1}{2}$ in. of his move to load and fire. He can fire and then move *or* move and then fire. He may stand still and fire twice.

To fire—the firer throws a dice and adds its score to his personal points value. Thus, a score of $5+1$ (personal point) will equal 6 which is a hit. The firer does *not* add his personal points value if:

(a)  The firer is himself under fire, without being behind cover.
(b)  If the target is behind cover.
(c)  If the firer is himself wounded.

When a man is hit, he throws a dice and adds his personal points value to the score.

Total 5 or 6—it is a minor wound of no effect.

Total 3 or 4—he is wounded so that he loses his personal points value and takes 3 in. off his move-distance. Figure laid down until next move.

Total 1 or 2—he is killed—figure laid down for rest of battle.

## Mêlées

Man fights man—each throwing a dice and adding personal points value and highest score wins. Loser throws hidden dice to see if he is (a) slightly wounded; (b) badly wounded and (c) killed. If (a) he fights on; (b) fights on less personal points value *and* 1 more point.

If there are 2 men fighting 1—the 2 each throw a dice plus

their personal points value—and the single man throws his dice plus personal points value. He can beat *one* of his two enemies if his total beats either of them. He is beaten if one of them totals more than him.

*Orders*
1. The bugler can be heard by anyone within 18 in. of him.
2. A shout can be heard by anyone within 9 in. of shouter.

1 means that the officer can give a *general* order (by *telling* his men the *same* order).

In 2 he can call out orders to anyone within 9 in. of him.

Otherwise, everyone writes down what he is going to do before each move.

| MOVE 1 | 2 | 3 | 4 | 5 | 6 |
|---|---|---|---|---|---|
| Run 12″ | Walk 9″ | Run 12″ | Walk 9″ | Run 12″ | Walk 9″ |

| MOVE 7 | 8 | 9 | 10 | 11 | 12 |
|---|---|---|---|---|---|
| Run 12″ | Stay | Stay | Stay | Run 12″ | |

Perhaps outside the scope of this book, as seemingly it encroaches upon the realms of psychology, man-management, human character and temperament, etc., is the question of which war-gamers are selected to represent the 'Lieutenant', the 'Sergeant' and the 'Strong man'. Because this 'Individualised War-gaming' really is an individual, personalised affair, the war-gamers are really 'on their own' on the war-games table, just as much as the riflemen and tirailleurs would have been in 1810. Similarly, the Lieutenant and the Sergeant really do have to command, make decisions, use their judgement and give orders that will bring purpose to whatever individual actions the solitary rifleman pursues through his own reasoning or personal initiative. When re-enacting this interesting little charade as a Club project, try to select someone with known powers of command as the Lieutenant; let the Sergeant be a steady, mature man (as indeed most Sergeants are!) and for the 'Strong man' select a war-gamer who is noted more for brawn than brains! It all makes for a highly interesting exercise in noting what makes a man tick!

When a large-scale campaign is being fought by a group of war-gamers it is possible to really go realistically into the field of logistics by appointing one man on each side as a Quarter-Master General. He will allocate supplies and keep his combat-colleagues informed as to the limitations imposed upon their tactical manœuvres by their supply-line. Roger Moores has worked out a really professional plan to cope with the situation in the horse-and-musket period.

1. Terms used:
    'Round'—One man's firing during one move in a battle.
    'Shot'—One gun's firing during one move.
    'Ration'—One man's feeding for one day.
    'Fodder'—One horse's feeding for one day.
    'Ammo'—'Rounds' and 'Shots'.
    'Food'—'Rations' and 'Fodders'.

2. Unit level Logistics. See Form 1.
(a) One man can carry 10 'rounds'.
(b) Crew limber can carry 10 'shots'.
(c) One man can carry 3 'rations'.
(d) One horse can carry 3 'fodders'.

FORM 1

Unit ...................................................................................................................

| Date | Strength | | | Rounds | Shots | Rations | Fodders |
|---|---|---|---|---|---|---|---|
| | Officers | Other Ranks | Horses | | | | |
| | | | | | | | |
| | | | | | | | |
| | | | | | | | |

Form 1 is used to record changes in strength, ammo and food. Where units are operating within half a day's march of supply units it is not necessary to maintain Form 1 (except for changes in unit strength) as units will be considered to be replenishing constantly from the Supply Unit. When the amount of food or ammo falls *below* the carrying capacity of men and animals, then Form 1 must be maintained, because Attrition rules will then apply.

'Living off the land' may be carried on by troops (including 'grazing' by horses) on food-producing areas on the map (marked in yellow) during the months April–September inclusive.

3. Attrition.

| After complete days | Without food | On half rations |
|---|---|---|
| 2 | ½ Speed. Deduct 1 from morale | No effect |
| 7 | No movement. Deduct 2 from morale ½ value in melee | ½ Speed. Deduct 1 from morale |
| 14 | Throw for desertion everyday, for each man. 1 means deserts. Otherwise, as for 7 days | No movement. ½ value in melee. Deduct 2 from morale |
| 28 | Throw for starvation. 1 means dies. 2 means deserts | Throw for desertion everyday. 1 means deserts |

Every 2 days of ½ rations must be cancelled out by 2 of full rations for normal conditions to return.

After 7 days, every 2 days of starvation must have been cancelled out by 2 of ½ rations or 1 of full rations for '½ rations' to start applying, and for 2 days of full rations for normal conditions to return.

Full rations on alternate days with no food equal ½ rations.

A well-fed horse, or one that has been on ½ rations for no longer than 7 days, is worth 20 rations when cut up and eaten. A starving horse or one on ½ rations over 7 days is worth 10 rations.

4. Supply Units (e.g. for 20 men, 3 Airfix 'Wagon train' wagons).

Each wagon carries:

100 Shots *or* 1,000 Rounds or equal proportions of the two mixed, or

2,000 Rations or 200 Fodders or equal proportions of the two mixed.

Recording of amounts in Supply Units is done on Form 2.

<table>
<tr><td colspan="5" align="right">FORM 2</td></tr>
<tr><td colspan="5">Supply Unit No.......................................................<br><br>No. of wagons...............................</td></tr>
<tr><td>Date</td><td>Rounds</td><td>Shot</td><td>Rations</td><td>Fodders</td></tr>
<tr><td></td><td></td><td></td><td></td><td></td></tr>
</table>

5. Supply Depots.

May be set up in towns anywhere at any time: see Form 3.

6. Rail Supply to Depots.

Supply Trains run at 200 miles per day from 'Factory' or 'Market Towns'. Maximum *single* line capacity—2 directions, is 10 trains each way.

A train may carry whatever 10 wagons can.

Trains may not be run on any section of line cut by the enemy.

<table>
<tr><td colspan="5" style="text-align:right">FORM 3<br><br>Depot at...................................................<br>(Place)<br><br></td></tr>
<tr><td>Date</td><td>Rounds</td><td>Shot</td><td>Rations</td><td>Fodders</td></tr>
<tr><td><br><br><br><br><br></td><td></td><td></td><td></td><td></td></tr>
</table>

7. Factories and Market Towns.

(a) (the simpler way). Let market towns and factories have unlimited capacity and govern supplies by the limitations of transport and the number of the factories and market towns.

(b) Allocate a 'rate' to factories and market towns. Use forms 4 and 5.

Market towns are usually in yellow areas on the map, and are marked with a green circle. Each accumulates Rations and Fodders from outlying areas during May, June, July, August and September only, at the rate of 2,000 rations and 200 fodders

<table>
<tr><td colspan="3" style="text-align:right">FORM 4<br><br>Market Town.............................<br></td><td colspan="3" style="text-align:right">FORM 5<br><br>Factory...................................<br></td></tr>
<tr><td rowspan="2">Date</td><td colspan="2">Amounts in Warehouse</td><td rowspan="2">Date</td><td colspan="2">Stock</td></tr>
<tr><td>Rations</td><td>Fodders</td><td>Rounds</td><td>Shot</td></tr>
<tr><td><br><br><br><br></td><td></td><td></td><td></td><td></td><td></td></tr>
</table>

per month (on the first day of each month). Factories produce shot at the rate of 50 per month, rounds at the rate of 1,000 per month, available on the first day of the month, all the year round. Mark factories on the map with red circles. They could also produce guns, muskets, etc.

These rules are designed for easy adaptation to the regiments of different war-gamers and are based on the '10-moves-a-day' type of game. Of course the proper limber/caisson capacities could easily and more realistically be used.

It might be found necessary to use another form for railway trains in transit.

The book keeping is quite simple. For instance when supplies are loaded on to wagons at a 'Market Town', simply enter the new total on Form 4 and on Form 2.

Other refinements to these rules could reflect the different staying powers of different types of troops. For example Confederates could last much longer on half rations than Union Troops, or French Napoleonic soldiers could 'live off the land' in areas that others could not.

# 10

# Strategic Napoleonic War-gaming
## as a Club Project

War-gamer John Bulgin uses the following strategic rules for his Napoleonic war-gaming—they appear to be an excellent means of employing a group of war-gamers in a campaign. John points out that the high hourly marching rate (4 m.p.h. for Infantry and 12 m.p.h. for Cavalry) is dictated by the size of counter he is using and the scale of map ($\frac{5}{8}$ in. diameter counter equals 4 miles on map). Obviously, the rate could be slowed down by using smaller counters, etc., or by making the 4-mile square a two-hour period.

The Daily Movement Sheet is the key to the system, enabling the two C.-in-C.s to supply the umpire with hour-by-hour information of the positions of all the units in his army.

### 1. *Maps*
Three identical maps of the strategic campaign area are required, each broken down into 4-mile squares and numbered so that each square may be identified by a number/letter code.

### 2. *Allocation of Maps*
The first map is given to the umpire and the others to the two C.-in-C.s.

### 3. *Object of Campaign*
Each side has three main cities, preferably at opposite ends of map. If the enemy succeeds by the rules described below in capturing all three cities, the troops on that side must surrender and the campaign is deemed over. These cities are marked on the maps with BLACK borders around their names.

### 4. *Identification of Units*
Each army is broken down into units representing Infantry, Cavalry or Artillery. Every unit is allocated a coloured counter, a different colour for each side. Each counter is numbered, i.e· British Army, red counters. Unit 1 = 12th Light Dragoons, etc.

Each side's C.-in-C. must make out a list denoting units and numbers.

### 5. *Defending Capital Cities*
If one city is captured, the defending army must intercept the enemy between first and second cities or successfully defend second city before advancing to retake first. The defending army cannot simply re-enter the first city whilst enemy attacks second.

### 6. *Moves*
(a) Each square on the map indicates an hour's march to an Infantry unit. All units may only march by way of main roads marked in broad red bands, and can only pass through squares that contain these roads.
(b) Units are not allowed to leave roads during strategic part of game whatever.
(c) Units may only march for a maximum period of eight hours in two four-hour stretches with an hour's rest in between. After this eight-hour march, they must rest for a minimum of twelve hours.
(d) Cavalry and Horse Artillery may move at the rate of three squares an hour for a maximum period of eight hours in the same fashion as Infantry.
(e) Moves for all arms at night are at half rate. Infantry=1 square every two hours. Cavalry and Horse Artillery=3 squares every two hours.

### 7. *Supply Towns and Supply Units*
(a) Various towns (marked by BLUE borders) are considered to be supply towns. Each unit requires supplies at the end of an eight-hour march when advancing. That is, every eight or twenty-four moves. Therefore, Army moves towards objectives along roads that link with supply towns.
(b) If an army column marches for eight hours and does not reach another supply town, it must HALT. It cannot move again except to retreat along its line of march unless it receives supplies from the town it has just left.

### *Example*
Presuming there are fourteen units in the army, each unit marches for four hours, rests for an hour, and marches for another four hours.

First unit leaves camp at 6 a.m., marches until 10 a.m., rests until 11 a.m., and then marches until 3 p.m. It must now rest for a minimum of twelve hours, so cannot march again before 3 a.m. the next day.

Fifth unit leaves camp at 10 a.m. and arrives at new camp at 7 p.m.

Fourteenth unit leaves camp at 7 p.m. and arrives at new camp at 4 a.m.

This puts the units in the following positions:

Units 1–7 at position 9.

Units 8 and 9 at position 8.

Units 10–12 at position 7.

Unit 13 at position 6.

Unit 14 at position 5.

This is assuming that dawn was at 6 a.m. and sunset at 8 p.m., the army marching at half moves, during the night. Therefore, presuming that the vanguard is three squares away from the next supply town and the rearguard four hours away from the last supply town, the supply unit may leave the last supply town at 4 a.m. Travelling at one square per hour (night half moves) until 6 a.m. it will arrive at position five at 7 a.m., position six at 7.30 a.m., position seven at 8 a.m., position eight at 8.30 a.m., and position nine at 9 a.m. It is supposed that supplies are simply dropped by supply unit as they go through so that no time is lost. When all units have been supplied, the army may start to march again (9 a.m.). The supply unit returns along the line of march to its town, arriving at 1 p.m.

*Note:* A supply unit may ride twelve hours before resting. Then must rest for twelve hours.

The vanguard arrives at its new supply town at 12 noon. All surplus supplies are now given up, i.e. vanguard has only used three hours' supplies and gives up remaining five, being re-supplied by the town. The next twelve hours is taken up in occupying the town and sleeping. The whole army moves into the town. Following this period, the army may march another eight hours, being supplied by the town.

8. *Line of March*

(a)  As the armies progress from square to square, a numbered sticker is put in the square behind the rearguard.

| 1 | 2 | 3 | 4 | 5 | 6 | 7 | 8 | 9 | 10 | 11 | 12 | X | X | X | X | X |
|---|---|---|---|---|---|---|---|---|----|----|----|---|---|---|---|---|

X = Units. Thus marking the route that the army has taken.

(b)  The line of march is the supply route followed by the supply units and they may *not* use any other routes whatsoever to replenish the army, or return to the base town.

(c)  If an army is defeated in battle, it may only retreat along its own line of march.

(d)  If the line of march is intercepted by the opposing army, say at position 17, as indicated by the numbered stickers, the intercepting army is given the date and time the enemy VANGUARD passed through that square as well as its direction of march by the umpire.

(e)  If the line of march intercepted by the enemy is between his forces and the latest supply town, it is considered that his supply route has been cut. In consequence, the force that has had its route cut will stop as usual after each of its units has completed eight hours' marching. Any units that have not reached a new supply town will await the supply unit. It will leave the last town as usual but is captured by the enemy. When the army realises that supplies are not forthcoming, it must retreat along its line of march until it reaches the square held by the enemy. This, of course, means that the force occupying the square has had considerable time to deploy its forces in the best defensive manner and in all probability occupies a strong position. A battle takes place. If the occupying side wins the other army retreats through or around its opponents, moving a further eight hours back along its line of march, regardless of being supplied. If the occupying side loses, it must move back along its line of march for eight hours in a similar manner.

The rule then is, when advancing supplies are needed. When retreating, they are not.

(f)  If an army retreats along its line of march and, having been re-supplied, wishes to advance again, it does not need to follow its original line of march.


### 9. *Forced Marches*

If a unit marches for ten hours in any one day this is termed a forced march. A unit can only march at this pace for two days in any week (a week consisting of five marching days, the remaining two being compulsory rest days regardless of forced marches or not.

### 10. *Cavalry Patrols*

Cavalry units may be detached from the main body to scout for the enemy. It must, of course, be remembered that they only travel at three squares per hour during the day or three squares per two hours at night, and can only travel for eight hours without supplies. If the army is stationary it does not use supplies at the same rate as on the march and can, therefore, replenish patrols if they return to camp within eight hours. If the army is stationary, no account need be taken of supplying it from the town because it may gather sufficient for its needs by foraging.

### 11. *Sources of Information*

When a supply town or capital is entered by the VANGUARD OF THE ARMY, a messenger is dispatched at the speed of cavalry to the opposing army. This horseman is allowed to forage and does not need supplies, but like everyone else, must not exceed eight hours' riding and twelve hours' rest. He travels by the most direct road route and informs the army of the date and time that the enemy entered the town he came from. By this means both sides have some idea of each other's whereabouts. Messengers cannot be intercepted.

### 12. *Battle*

When two units (one of each side) move into the square an interception is deemed to have taken place. Whilst these units manœuvre on the war-games table a messenger is sent to all other units in turn, who, when knowing of the interception, converge on the battle square. A battle under tactical rules then takes place. The losing side retreats for eight hours along its line of march.

Example :-

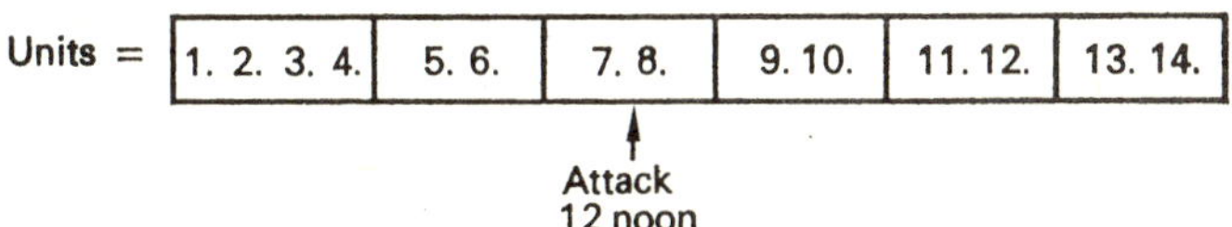


Whilst seven and eight deploy, horse messengers are sent to five, six, nine and ten. First messenger arrives at five and six at 12.20 p.m. which are both Infantry units. Marching at one square per hour they arrive in battle square at 1.20 p.m. or

F

during the 2nd game-move. (12–1 p.m. 1st game-move. 1–2 p.m. 2nd game-move.) One side takes half hour of game-move to complete their movements, the other side the remaining half. Therefore, fifth and sixth Infantry units arrive after enemy's second move and before British second move. Ninth and tenth units are visited by second messenger at 12.20 p.m. similarly. Both are Cavalry units and therefore arrive at battle square at approximately 1 p.m. or half an hour earlier than Infantry. Unit 1, 2, 3, 4 (all infantry) arrive at 2 p.m. together with Infantry 11, 12. Cavalry 13 arrives at 2 p.m. and Infantry 14 at 4 p.m.

### 13. *Duties of C.-in-C.s*

Each C.-in-C. has one map. On this he plots by means of the coloured counters the position of his units and by using the rules outlined above endeavours to capture the enemy's capitals. He plots his own cavalry patrols and when messengers reach him (controlled by the umpire) he may plot the last known position of the enemy. He must supply the umpire with details of all his troop movements from midnight to midnight on the Daily Movement Sheet—see example. This indicates each unit's position once per hour.

### 14. *Duties of Umpire*

Possesses third copy of map. Is fed written information from C.-in-C.s and plots *both* armies on map. Organises travel of messengers across map when troops enter supply towns, etc., and tells C.-in-C. if his army intercepts opponent's line of march. Generally co-ordinates movement and supplies necessary information, also checking that movements are legal. Plots both armies' lines of march.

TABLE OF DAWN AND DUSK TIMES FOR THE COMPUTATION OF
NIGHT MOVES

|  | *Dawn* | *Dusk* |
|---|---|---|
| January | 7.0 a.m. | 4.0 p.m. |
| February | 7.0 a.m. | 5.0 p.m. |
| March | 6.0 a.m. | 6.0 p.m. |
| April | 5.0 a.m. | 7.0 p.m. |
| May | 4.0 a.m. | 8.0 p.m. |
| June | 3.0 a.m. | 9.0 p.m. |
| July | 3.0 a.m. | 9.0 p.m. |

| | | |
|---|---|---|
| August | 4.0 a.m. | 8.0 p.m. |
| September | 4.0 a.m. | 7.0 p.m. |
| October | 5.0 a.m. | 6.0 p.m. |
| November | 6.0 a.m. | 5.0 p.m. |
| December | 7.0 a.m. | 4.0 p.m. |

Daily movement sheet

| Time | 1 | 2 | 3 | 4 | 5 | 6 | 7 | 8 | 9 | 10 | 11 | 12 | 13 | 14 | 15 |
|---|---|---|---|---|---|---|---|---|---|---|---|---|---|---|---|
| 12M | | | | | | | | | | | | | | | |
| 1 | | | | | | | | | | | | | | | |
| 2 | | | | | | | | | | | | | | | |
| 3 | | | | | | | | | | | | | | | |
| 4 | | | | | | | | | | | | | | | |
| 5 | | | | | | | | | | | | | | | |
| 6 | D1 | D1 | D1 | | | | | | | | | | | | |
| 7 | D2 | D1 | D1 | | | | | | | | | | | | |
| 8 | D3 | D2 | D1 | | | | | | | | | | | | |
| 9 | D4 | D3 | D2 | | | | | | | | | | | | |
| 10 | D4 | D4 | D3 | | | | | | | | | | | | |
| 11 | D5 | D4 | D4 | | | | | | | | | | | | |
| 12N | D6 | D5 | D4 | | | | | | | | | | | | |
| 1 | D7 | D6 | D5 | | | | | | | | | | | | |
| 2 | D8 | D7 | D6 | | | | | | | | | | | | |
| 3 | " | D8 | D7 | | | | | | | | | | | | |
| 4 | | " | D8 | | | | | | | | | | | | |
| 5 | | | | | | | | | | | | | | | |
| 6 | | | | | | | | | | | | | | | |
| 7 | | | | | | | | | | | | | | | |
| 8 | | | | | | | | | | | | | | | |
| 9 | | | | | | | | | | | | | | | |
| 10 | | | | | | | | | | | | | | | |
| 11 | | | | | | | | | | | | | | | |

UNITS

Map positions

DATE: 10th August 1814

# Re-fighting Real-life Campaigns

# Re-fighting the Franco-Prussian War

To try and reverse historic decisions of actual battles fought long ago on well-known fields is a most attractive prospect. But it is not an easy matter to simulate the ebb and flow of the campaign as it actually occurred. It only requires an army to win a battle on the table-top which they lost in reality to throw the whole scheme out of joint, because the original sequence of events has been interrupted and the army which advanced in real life will now be thrown back to a site perhaps never occupied by them during the actual war. The Author has employed, with reasonable success, two methods of reproducing actual campaigns and readers will no doubt be able to perfect either or both of these methods to suit themselves.

In the first instance, it was an attempt to reproduce the Franco-Prussian War of 1870. It will be recalled that this was a war of continuous Prussian success with the French armies reeling back before their onslaught without a single noteworthy forward movement. To re-fight such a predestined affair would be dull indeed; at the same time, if the French are to win on the table-top where they lost in 1870, this Prussian forward movement would be altered most unrealistically.

One overwhelming advantage of re-fighting an actual campaign lies in the fact that the battles are known and maps showing troop dispositions and terrain features are readily available. Taking the majority of the principal battles of the Franco-Prussian War, they were arranged as follows:

| | | |
|---|---|---|
| Worth | Spicheren | Weissenburg |
| Colombey | Nouilly | Borny |
| Rezonville | Vionville | Mars-La-Tour |
| Gravelotte (South) | St Privat | Gravelotte (North) |
| Beaumont | Sedan | Bazeilles |

Both the Prussians and the French are each divided into three separate armies (as they fight separately, this can be covered by using all your troops for each army). Both commanders number their armies 1, 2 and 3 or A, B and C. Now comes the crux of the matter, the steps taken to cope with a reversal of history—a Prussian defeat on the table-top. Take the three centre battles Rezonville, Vionville, and Mars-La-Tour. Each general will nominate one of his armies to fight an opponent's army on each of these battlefields. By choosing the 'middle' battles of the war, we are now in a position to have both the French and the Prussians moving forward or retreating in accordance with the results of the battles on the table-top.

Copy, in reasonably diagrammatic fashion, the maps of these three battlefields. Right around each battlefield draw in a continuation of the countryside, the roads, the rivers, etc., so that this border is equivalent on each side of the battlefield to a scaled war-games table. Let us take the battlefield of Vionville as an example. From this battlefield continue roads leading to Nouilly in the North; to St Privat in the South; to Rezonville in the West and Mars-Le-Tour in the East.

Now fight your battle with your two nominated armies in your usual fashion and using your normal rules. On the conclusion of the battle the loser may take one of the following courses of action:

1. He may retreat his force along the road to the battlefield immediately behind him.
2. He may retreat his force to Rezonville or to Mars-La-Tour on his East or Western flank.
3. If the situation permits, the loser may move the remnants of his defeated army into the square immediately behind his battlefield on the map. In this way he may use some of his forces to cover the retreat of the remainder. If he does this, a further battle will take place in this square with both sides being permitted to lay down up to within 6 in. of the centre line. Each side will lose an agreed percentage of losses from the last battle, before the second battle commences.

If an army manages to get a force of not less than one cavalry regiment across the road leading to the battlefield immediately behind the one on which the combat has just taken place, then the loser has the following alternatives:

1. He may attempt to retreat along another road to an alternative battlefield.
2. He may fight his way through the intercepting force. In this event, the retreating force will throw an individual dice for

each man to the same number as there are in the intercepting force *plus* the numbers of *all* troops of the intercepter's side within one cavalry move of the intercepter's commander. The intercepting force throws for half their number.

Having done this, the remainder of the retreating force are deemed to have broken through and will, on arrival at the new battlefield, be considered to be fatigued and will only move half moves for the first two game-moves in the next battle.

# 12

# Re-fighting the Peninsular War

Throughout the summer of 1966 the town of Hastings in Sussex celebrated the 900th anniversary of the Battle of Hastings, fought on the 14th October, 1066. Part of these events consisted of the public exhibition of a large model of the battle-field and battle itself. Constructed by Donald Featherstone, the battlefield was 18 ft. by 16 ft. in size, built up on twelve 6 ft. by 4 ft. baseboards. More than 3,000 model soldiers were positioned on the model to depict a vital stage in its course. These figures were 30 mm flats painted by Tony Bath. (In 1967 the Corporation of Hastings decided to alter the model and it now depicts some sort of mediaeval conflict between hordes of converted Airfix figures milling around in front of gaily striped tented camps inhabited by priests, women and the like. Highly colourful, it bears only the most fanciful resemblance to the Battle of Hastings.)

On the first Saturday of each month throughout 1966, the author, together with Tony Bath and a group of volunteer war-gamers, removed the four centre sections from the model and transported them to a nearby hall. Here the figures were laid out in the positions they occupied at the start of the actual battle and a reconstruction took place, accompanied by a spoken narrative. Every feature of the battle was performed in its correct chronological order, casualties being removed in scaled-down proportions to those that actually occurred in 1066. The battle culminated in Harold's death and the fall of the standard.

This is one way of re-fighting a battle. It is a staged, timed and accurately carried out version of the events that actually took place but using perhaps one figure to represent five or ten of the men who took part in the original battle. This is not a war game as it has no competitive element nor is the result in any doubt.

At the same time, the simulation of any actual battle, whilst holding an immense fascination and interest, is far from easy and will frequently only bear the most coincidental resemblance to the events that actually took place in the perhaps distant past. The main factor mitigating against a realistic representation lies in the fact that inevitably the numbers involved on the table-top must be a ridiculously scaled-down proportion of the real numbers—a fact which immediately and inevitably detracts from any idea of realism or an accurate result. The rules under which the battle is re-fought must be carefully formulated so as to accurately and realistically reflect the specific features of the time so that the weapons and their effect together with national or local peculiarities of the troops involved are brought into prominence.

Nevertheless, it is a project worth attempting as was seen when the author and Neville Dickinson (of *Miniature Figurines*) attempted to re-fight the Peninsular War. Using Jac Weller's admirable book *Wellington in the Peninsula* as a guide, maps were drawn from those given in the book that were capable of being re-created on the table-top.

If the war game is to represent the actual battle in anything more than its name, the tactics and events of that battle have to be reproduced to a certain extent. The war-gamer commanding the side that was victorious in real life has an advantage through this method whilst the commander of the defeated side, with hindsight, knows what not to do on this occasion.

When drawing the terrain map it was considered that if the battle contained more than one 'part' (i.e. two or more events of a major nature) then it could either be fought by splitting the table-top terrain so that each part takes the form of a separate battle or by 'coming into the battle' just prior to what is considered to be the major or most interesting facet of the engagement. For example when fighting Rolica, it was decided that the second part of the battle, when the French under Delaborde had fallen back from their first position because of a British out-flanking movement and had taken up a second and stronger position on a ridge some two miles further back, was a more interesting proposition.

In order to obtain a reasonably accurate scaled-down force, it was decided to let one man on the war-games table equal forty men in the actual battle. For Rolica, this gave the British a force of about 350 men and the French 115 men, balanced by the French having the advantage of being in a very strong position and also possessing more cavalry than the British.

The Peninsular War is a very suitable campaign to re-create

because it is possible to follow the sequence of battles as they actually transpired because Wellington won all his battles in the Peninsula. On a war-games table considerable difficulty is encountered if the sequence is broken by the enemy daring to be successful in a war game when they were defeated in real life. The inevitable result will be that instead of the armies progressing conveniently on to the next battlefield as they did in the past then, in the Peninsular War re-enactment, the British will probably have retreated to an unknown battlefield upon which no actual engagement ever took place.

This can be countered by allocating 2 points to the victors, 1 point for a drawn battle and additional points if any specific objectives are gained or held by one or other of the sides. For example, as can be seen in the description of the re-fought Battle of Rolica, it was considered necessary for the smaller French force in their strong hilltop position to hold off the considerably stronger British force and yet still be able to make their getaway before the arrival on their flank of the strong British detachment under Ferguson. In the event, the French were able to do this and it was decided that the British had gained 2 points for the victory whilst the French had gained 1 point for holding off the British for 14 game-moves.

The original Battle of Rolica was fought on the 17th August, 1808, when a French force under General Delaborde were in a strong position before the village of Rolica in the centre of a horseshoe of hills surrounding irregular and broken terrain. Wellington was advancing and endeavouring to engage Delaborde before he managed to join forces with another French army under Loison, half a day's march to the east. Wellesley (at this stage it is better to so title the Duke of Wellington) divided his force into three columns, the most powerful of which under his direct command moved down the centre of the valley towards the French position at Rolica. It consisted of $3\frac{1}{2}$ Brigades of British infantry, a weak battalion of Portuguese Cacadores, a few British and Portuguese cavalry and 12 guns, in all about 9,000 men. Far in advance of the main British force to the right and west, a flanking column of about 1,400 Portuguese under Trant hastened through the hills as one half of a pincer to encircle the French, the other half of this same movement was a far more powerful column to the left and east, commanded by Ferguson consisting of 2 Brigades of Infantry, half a brigade of Fane's riflemen, a detachment of 40 cavalry and 6 guns. This column, about 4,500 strong, also had the task of protecting the flank should the French under Loison attack.

After some skirmishing between the French tirailleurs and Wellesley's light companies, Delaborde retired behind his cavalry screen to a second position two miles further south which he had chosen some days earlier. The French occupied a section about three-quarters of a mile wide of a horseshoe of hills, with each flank resting on deep breaches in the hills worn by winter streams. The front of the position appeared too steep to climb, except where four boulder-choked gullies reached back into the ridge. Along the summit of the ridge a natural dyke stretched which gave the French a fine natural breastwork. Their cannon from the top of the ridge could sweep the valley through which the British had to advance. Their force consisted of 5 battalions of Infantry; 3 squadrons of Cavalry; and 5 guns —a total of about 4,400 men.

Wellesley's main column advanced within artillery range and deployed into line of battle whilst spirited skirmishing between the light companies and the French tirailleurs at the foot of the slope commenced at once. Sir Arthur did not intend to press his attack until his flanking forces were sufficiently far advanced to aid the main attack. Meanwhile he kept up pressure with his skirmishers and artillery to make a French withdrawal difficult. After Colonel Lake had inadvertently got a small and disordered British column up one of the boulder-stewn gullies which ran deep into the French position, being killed himself and taking heavy losses, Wellesley realised that severe fighting was inevitable and ordered an immediate general attack. The whole British line moved forward up the four gullies and the steep slopes between them. They were unable to take advantage of their numerical superiority nor could they use any definite formation.

Units found ways which enabled them to reach the summit without serious opposition and soon there was a continuous line of British infantry formed along the western half of the ridge itself and fighting of a severe nature took place.

Once the French position was reached, Delaborde accepted defeat but minimised it by skilfully withdrawing his four infantry battalions alternately two at a time, covered by his cavalry. Wellington did not press his advantage because he had heard that Loison's forces were only five miles away. Wellesley's army was three times as strong as Delaborde's, but the numbers actually engaged were about equal. The French lost about 700 compared to 485 British and they lost three-fifths of their artillery.

### THE BATTLE OF ROLICA FOUGHT AS A WAR GAME

This battle was entered at the stage where Delaborde had retreated back and was strongly in position on the top of the ridge approached by four boulder-strewn gullies and very steep slopes. The terrain was laid out on an 8 ft by 4 ft. table as shown in the accompanying map.

From his force of 350 men, Wellesley had to detach two flanking columns, leaving him with a force of about 200 men, all infantry plus three guns. The French, behind a convenient rocky breastwork, had 115 infantry, 15 cavalry and 1 gun.

Delaborde was aware that a strong flanking column was approaching him from the east but he did not know when to expect its arrival; on the other hand, neither did Wellesley know when his column would arrive, being unaware of any difficulties they might encounter en route. To represent this on the war-games table, an impartial umpire threw 3 dice and added the totals together; this total represented the number of game-moves at which Ferguson's flanking column would arrive on Delaborde's right flank. The proviso was made that Delaborde would be told 3 moves before the column actually arrived as it was considered likely that he would have seen them approaching from his lofty position.

The battle opened by a strong British column advancing up the westernmost gully with a smaller column advancing up the eastern gully. The centre/eastern gully was rapidly traversed by a force of riflemen whilst the centre/western gully held a reserve force of 2 battalions of infantry whilst a gun was thrown forward halfway up the gully and brought into action. A gun also accompanied the riflemen up the centre/eastern gully.

When within range a sharp fire fight ensued between the two eastern columns and the far western column. The latter struggled to deploy on emerging from the top of the gully, their numbers impeding them and making this manœuvre difficult. They were charged by the French cavalry and similarly engaged by a battalion of French infantry, the resulting mêlée continuing for number of game-moves. The British eastern flank came under fire from the French artillery and the riflemen, emerging from the top of the gully, deployed into skirmishing lines and engaged similar lines of French tirailleurs. After about 4 moves the French gun was knocked out by artillery fire, a very severe blow to the French as they now had no artillery whatsoever. A French

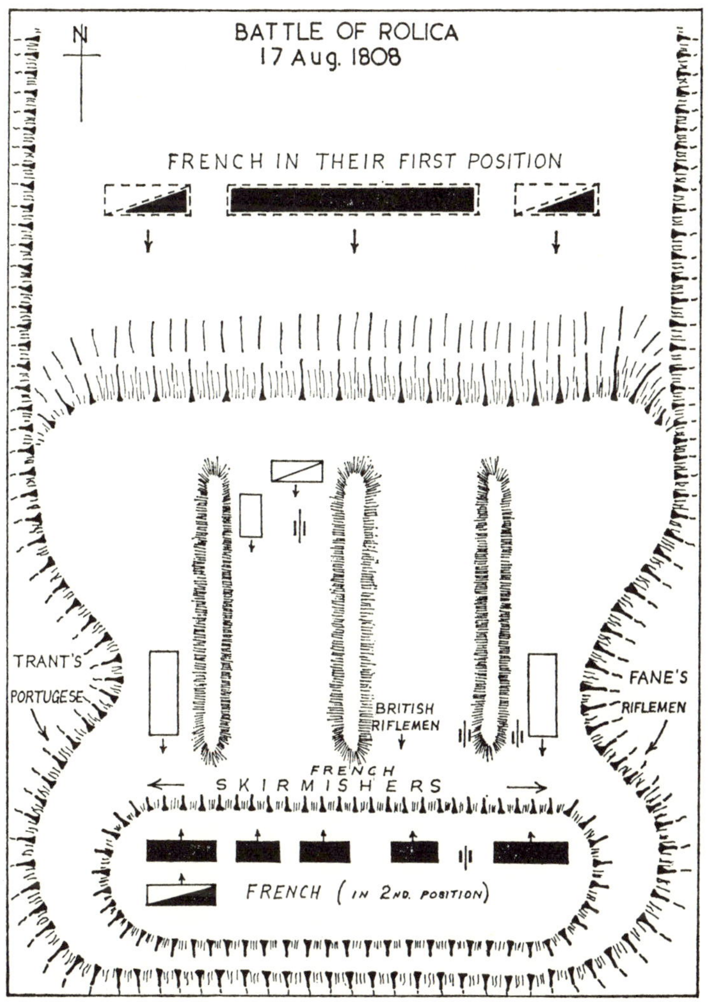

infantry regiment, taking severe losses, found their morale to be insufficient to stand the strain and retreated back out of line.

Delaborde was becoming a little apprehensive as the battle went on, wondering when Ferguson's flanking column was going to approach. It was coming from the British left on which flank the British forces were doing the most damage and making the greatest advances. On the 11th game-move, Delaborde could stand the strain no longer and disengaged his cavalry and began a withdrawal. He was close enough to the rear end of the table to be able to successfully evacuate his forces without them being engaged by the British, who lacked cavalry.

It transpired that the 3 dice thrown to indicate the time of arrival of the flanking force had turned up two 6's and a 4 so that Delaborde had two more moves before he would have been told that the British were approaching and then three more moves to get off the table. Under the circumstances it was considered that although this was undoubtedly a British victory, the French had earned themselves 1 point so that the final score was Wellesley 2 points, Delaborde 1 point.

Chronologically, the next engagement in the Peninsular War was the Battle of Vimiero followed by Corunna and so on. These battles and war were fought out until Fuentes de Onoro took place on a table-top terrain. Just as had occurred in every other battle of the campaign, the British again won, showing that the rules had realistically represented the superiority of British line over French column. The war-gamer handling the French appeared to be a little discouraged and, considering that he was not 'in with a chance', the campaign was terminated at this point.

# 13

## Re-fighting an Actual Battle
### (Spicheren, Franco-Prussian War, August 1870)

Perhaps the most interesting method of re-fighting an actual battle is to draw a simplified map of the actual battle area, assemble two armies and then re-fight the battle on the same terrain but allowing each commander to carry on as he wishes. The commander who was defeated on the actual day can with hindsight reverse the decision by avoiding the errors his historical predecessor made and at the same time he can carry out choice tactical schemes of his own. It is interesting to consider that two major battles of the 19th century, Waterloo 1815 and Gettysburg 1863, have been fought on a number of occasions by the author and his associates and rarely have the original winners been successful!

The Battle of Spicheren on the 6th August, 1870, was one of the earlier engagements of the Franco-Prussian War. The terrain over which the battle was fought is most interesting although the battle itself could with some justification be termed a 'dog's dinner'. The French General Frossard had fallen back with his 2nd Corps to the Spicheren position amid the heights above Saarbruck. The advancing Prussian First Army came under fire from these heights but believed them only to be occupied by a French rearguard so an immediate attack was ordered. As no battle had been expected on that day the march of the Prussian troops was badly arranged so that regiments were only coming up by unprescribed routes and arriving one after the other at different hours. This meant that the strong French position was attacked in the beginning by only one Prussian brigade which was gradually reinforced as Prussian troops arrived and were flung into the battle.

The position occupied by Frossard's Corps was very strong. From the centre projected the Rothe Berg, precipitous and almost inaccessible cliffs, and the slopes on either side were densely wooded. On the French left the massive buildings of Stiring Wendel ironworks furnished a separate defensive position. Just beyond Stiring and extending to the Saar River

G

was the impenetrable forest of Saarbruck. Although Frossard was holding this position with only one Corps there were no less than four other French Corps behind him at distances varying from 9 to 19 miles so that it was within the power of the French to collect five Corps to support Frossard in the Spicheren position. But, true to the French handling of this war, nothing was done.

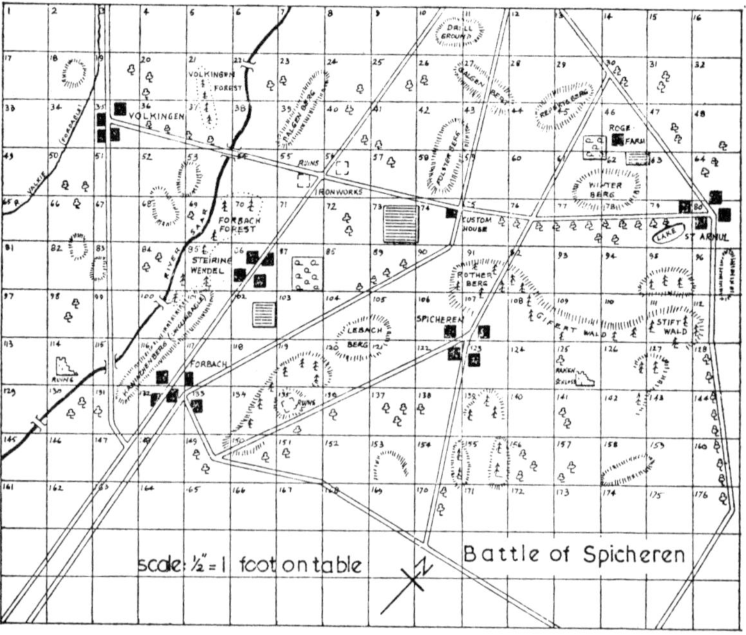


Fighting bravely and constantly being reinforced although in a piecemeal fashion, the Prussians steadily advanced, throwing the French out from one strong position after another until by nightfall they were falling back over the whole plateau.

A map was drawn to a scale of $\frac{1}{2}$ in. = 12 in. on the war-games table; it was sixteen 1 in. squares wide and thirteen 1 in. squares deep. At the stated scale it represented a block of war-games tables 4 by 5 roughly.

*Forces*
Each side has the following:

3 Light Regiments (Chasseurs or Jagers respectively).
18 Line Regiments.
3 Cavalry Regiments.
6 Guns.


*Original dispositions*
Each side may divide their forces into whatever strengths they
desire and they must initially be nominated to the following
Assembly Areas:

*Prussian*—Area No. 1—Squares 17 18 19 and 20 (Flar'k area)
                      2          21 22 23 and 24
                      3          25 26 27 and 28
                      4          29 30 31 and 32
*French*—Area No.  1         49 50 51 and 52 (Flank area)
                      2          53 54 55 and 56
                      3          57 58 59 and 60
                      4          61 62 63 and 64

Both sides *must* assemble in at least *three* of the four areas in
reasonable strength.


*Laying-out*
For the first battle it has to be decided if both sides are advancing to find each other *or* are the French setting up defensive
positions? It must be borne in mind that the Prussians, generally
speaking, are advancing.

If it is the former then it is an Encounter Battle and it must
be decided if both sides lay down on the baseline, 12 in. in
from the baseline or 12 in. from the curtain across the table
centre.

If it is the second choice then the French lay down initially in
defensive positions and the Prussians advance from the Base
Line.


*Set-up*
Dice throws decided that the first battle takes place in Area No.
4 which means that the advancing Prussians move forward so
that their baseline is the top of squares 45, 46, 47 and 48. The
battle terrain consists of *Squares* 45 46 47 48
                                    61 62 63 64
                  Top ½ of 77 78 79 80
Dice throws also decide that the French are setting up

defensive positions, so in this battle the French lay down first not further forward than Roge Farm area.

*Further developments*
If Prussians win this battle they may select the next battle terrain in the following manner:

1. The losing French may withdraw back (straight or in diagonal) any distance they choose but not more than $2\frac{1}{2}$ squares measuring from their baseline (top of bottom half of Squares 77, 78, 79 and 80).

2. The Prussian transparent template is then placed on the map in any logical position adjacent to this battle area but not further forward than with its front edge on top of squares 125, 126, 127 and 128 or similar line. The area covered is then the area of the next battle.

3. The Prussians who have just won may move forward to occupy the ground just captured and remain. The remainder of the Prussian forces may then select their next battle area— this may be *any* $8\frac{1}{2}$ squares covered by the template *but* no further forward than line 113 to 128 and dependent upon any French resistance in any of the areas over which this forward move might take place. (This new area may overlap the area already used if desired, it does not have to strictly conform to the original assembly areas.)

4. If the French win first battle: The defeated Prussians may retreat up to 2 backward or diagonal moves and the French may move forward to again contact them.

If the French do not wish to do this, then the Prussians who have lost will retreat as above and the remainder of the Prussian forces may select their next battle site as explained above. This will be the pattern after each battle. The specific situation will decide whether the next battle is to be encounter or whatever.

*Map moves*
Map moves of flanking forces will be at the following rates:

(a) *Ordinary marching:* Line Infantry and Artillery—$\frac{3}{4}$ in. per *two* game-moves.

Light Infantry—1 in. per *two* game-moves.

Cavalry—$1\frac{3}{4}$ in. per *two* game-moves.

(b) *Forced marching :*     Line Infantry and Artillery—1 in. per
                            *two* game-moves.
                            Light Infantry—1½ in. per *two* game-
                            moves.
                            Cavalry—2¼ in. per *two* game-moves.
Plus 25% bonus on roads.

Troops carrying out a forced march will not be able to move above half speed on the table.

Other than map-moving for relatively unopposed flanking forces, moves on maps for other troops will be largely a matter of expediency to give a good situation!

*Attrition*

It is not intended that there shall be any losses in man-power carried on from one battle to another as a result of casualties. Thus, after Battle No. 1 both forces will regroup as outlined above in their original strengths.

This was a highly successful campaign which took three or four battles to complete. The original lay-out meant that the French (at the south-east or lower part of the map) had a force on their right wing with no enemy facing them whilst the Prussians (at the top of the map) were in a similar position on their right wing. The Prussians utilised this spare force to come down the tree-lined road that led from square number 3 down to square number 147. They then turned up to Forbach so forcing a French Corps to turn on its axis and fight them facing in the opposite direction to the rest of the French Army. The Prussians eventually won this campaign—but the Prussians always seem to win our war games against the French!

War-gaming offers many intriguing facets and diversions because it is an attempt to simulate the reactions of human beings under stress. All rules pay lip-service to the conviction that they are going a certain way towards this simulation but the end result is often disappointingly chess-like. Of course a great deal depends upon the imagination. For example, in the Horse-and-Musket period of warfare, in addition to Man's natural fear of death or mutilation, the features of smoke and noise greatly increased his apprehension. Smoke and noise are noticeably absent from war games and can only be present in the imagination of the war-gamer moving his carefully painted armies over a terrain that looks far from natural.

It might look good on the war-games table to see the red-trousered French and the blue-uniformed Prussians moving steadily forward in the serried ranks that facilitate the easy firing of war-games volleys—but is it realistic or natural? In real warfare battalions and regiments withdraw or are beaten because of:

(a)  The total casualties they have suffered.
(b)  The number of officers they have lost.
(c)  The numerical odds against them.
(d)  The weight of fire they sustain plus the varying effect of that fire being received frontally or from flank or rear.

This late 19th century Horse-and-Musket battle of Spicheren gave the opportunity of testing a method of tackling the factor (d) from the list given above. The method was based on the question whether or not soldiers would continue to move forward in an erect position when under fire. It would seem more natural to assume that when coming under fire troops would do one of three things—they would either:

(a)  Go to ground.
(b)  Slow down.
(c)  Move faster.

It is also likely that those reactions would be affected by the type of troops concerned. It is reasonable to anticipate that disciplined troops would conduct themselves better than those lacking the same degree of discipline—this leads on to the assumption that a Guard Regiment will be steadier than a Line unit and both will be better fitted to withstand fire than a Militia force. Native troops might not be quite so steady although this depends to a certain degree upon the presence of their known white officers. A degree of uncertainty is introduced here when one thinks of such units as the Gurkhas. Irregular troops such as tribesmen or guerillas are likely to be the least steady under fire and to show an impulsive reaction. On the other hand one must take into account the incredible bravery and determination to move forward under fire of such Irregulars as the Fuzzy-Wuzzies and the Zulus (although the latter were soldiers in their own right). To take the matter even further, national characteristics can be considered—for example if re-fighting the Libyan campaign it would be necessary to evaluate the known dash of the Australian infantry against the notable reluctance of their Italian counterpart.

In what can be termed a very modest attempt to represent all these factors, it was decided in this game that all infantry should go to ground immediately on coming under fire—in other words, they would lay down. At this stage future action had to be decided—they had three choices:

1. If they stay down they suffer no casualties but they cannot return fire.
2. Still down, they can return fire but take *half* casualties.
3. They can rise to their feet and move their full distance *or* move and fire with half effect *or* charge move to make contact with the enemy.

To decide which of these choices were made it was necessary to decide the morale of each group of men whenever fire was received. This was done by throwing a dice for each group under fire and deducting from it as follows:

> For *any* losses − 1.
> If reduced to half normal officer strength − 1.
> If the opposition outnumbers them approximately two-to-one − 1.
> If they are under artillery fire − 1.
> If they are under fire from flank or rear − 1.

To be able to rise to their feet and move or fire, the group needs to make the Fire-Morale total of 3.

To stay down with the ability to return fire but taking half casualties, the group needs to make a Fire-Morale total of 2.

Staying down without casualties and unable to return fire is the normal basic reaction and is what occurs if the group concerned are unable to make the required Fire-Morale totals.

In this particular game, troops under fire in this manner had the following firing ability:

1 dice is thrown per volley of 5 men.

At a range of 24 in. and under—$\frac{1}{8}$th of the dice score counts as enemy casualties.

At a range of 15 in. and under—$\frac{1}{4}$ of the dice score counts as enemy casualties.

At a range of 6 in. and under—$\frac{1}{2}$ of the dice score counts as enemy casualties.

Add 2 per dice if the target is in close order; deduct 1 per dice if the target is behind hard cover.

When men are firing from behind hard cover such as a wall. or a house or if they are not under fire then the value of their volleys is increased to $\frac{1}{4}$ score, $\frac{1}{2}$ score and $\frac{3}{4}$ score respectively.

Saving throws were allowed for all casualties. When a man was laying down then he was saved by a throw of 4, 5 or 6. When a man was behind cover he was saved by a throw of 2, 3, 4, 5 or 6. When a man was standing he was saved by a throw of 3, 4, 5 or 6. The apparent anomaly in the saving throw between a man laying down and a man standing occurs because it was decided that a man laying down would be most likely to be hit in the head with fatal results.

It is difficult to assess the true worth of any method on a single testing, nevertheless this method appears to hold promise and to be worthy of further trial. It would appear to produce the effect of a lower casualty rate but the same end results because units are realistically falling back, going to ground and staying there or generally reacting as men may quite reasonably do under stress in spite of being given orders to go forward. For the war-gamer who, like the author, prefers to see a table-top battlefield full of troops at the end of the battle and the result attained by tactical manœuvring, then this method may have much to offer.

SECTION FOUR

# Campaigns Through the Ages

# 14

## The Viking Raid

It would be difficult to select from any period of military history one particular warrior who stands out because of his warlike qualities, proficiency with weapons and his victories. However, if one is considering the fearful effect he had upon his opponents and the dread in which they held him, then it is not necessary to look very much further than the Viking from Scandinavia who ravaged England's shores in the 9th and 10th centuries.

Warning beacons flamed on the headlands at the first sight of the square sails and low black hulls of the longships and sent panic-stricken villagers fleeing for safety to some inland town. These invaders were not invincible, sometimes they were thrown back by superior numbers or by villagers turned into warriors through sheer fear and desperation. Whenever possible they avoided pitched battles—their main purpose was to loot and then retire to their ships with plunder and captives. The best seamen in the world, the Vikings were able to strike wherever they chose. Able to swoop down wherever the water was deep enough to float their longships, they were away with prisoners and booty, leaving only burning houses and scattered bodies to greet the forces hastily assembled to beat them off. Sometimes they swept far inland on stolen horses to descend like devils from hell upon villages so far inland that such attacks came like bolts from the blue.

Completely lacking in national feeling, most of the time the strongest prevailed in the Viking world where the only loyalties might be to kinsmen or neighbours. Their main weapons were swords, spears, battleaxes plus bows-and-arrows; they wore iron helmets sometimes fitted with horns or raven's wings and shirts of leather with plates or scales of iron sewn on them. The Vikings carried round wooden shields with an iron rim and boss, usually they were painted in intricate designs.

The most dreaded of all the Vikings were the Berserks who, disdaining all protective armour, fought without a serk or shirt of mail. Berserkers were warriors who, in the stress of battle, were likely to throw themselves into some kind of a fit so that they howled, foamed at the mouth and bit the edges of their shields until they were in such a frenzy that they had convinced not only their enemies but themselves that they were invincible. This fighting madness enabled them to perform feats of daring and endurance far beyond the normal capabilities of mankind.

Such colourful and fierce fighting men lend themselves naturally to war games and the simulation of their feats. On the other hand it might be claimed that the meagre opposition they encountered destroys any hope of a good war game before the table-top terrain has even been laid. This is readily countered by methods and rules that cause the Vikings to be affected by the many varied factors which might well count against such men in real life—invariably they will be inferior in number to the less fierce and not so well armed British, they will be fighting over unfamiliar ground well known to their opponents, whilst their greed for plunder might lead them to over-reach themselves or Nature might come to the aid of the British and aid the downfall of the invaders by sudden changes in the weather.

It has already been said that the Vikings came to loot and get away with the maximum plunder and the minimum loss. This means the campaign is an exercise in bringing the Vikings to battle—such a campaign holds out great scope for manoeuvring in an unusual and interesting manner. A number of interesting battles can be suggested ranging from a single boat with perhaps 20 or 30 Vikings as its crew, landing on a beach and being opposed by two or three times that number of villagers. Or the map be so drawn as to present a coastline with two villages fairly close to each other with the complication that one village is willing to defend itself whilst the other is frightened and its men may submit or run away.

Then there could be a map with two beaches divided by a mouth of a river and with a village on either side. Here the problem will be for the British to decide where the invaders are going to land and for the Vikings to attempt to out-smart them. Back in the 8th and 9th centuries, the usual procedure of the Vikings was to seize an off-shore island so as to secure an easily defensible base for future operations. This could be simulated as a war-games campaign by using one of the Ordnance Survey maps such as 'Britain in the Dark Ages' which covers the period between the end of Roman rule and the time of King Alfred (approximately A.D. 410 to 870) or the map of 'Ancient Britain' covering the period of earliest times to A.D. 1066 The

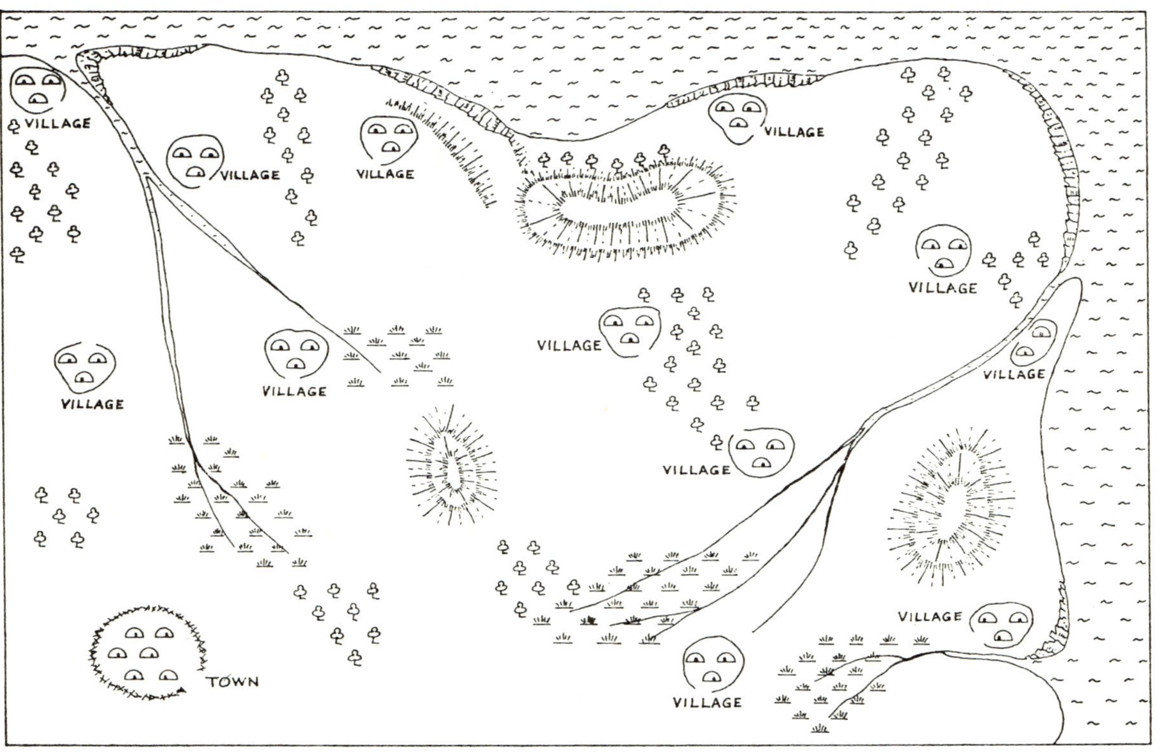

VILLAGE
VILLAGE
VILLAGE
VILLAGE
VILLAGE
VILLAGE
VILLAGE
VILLAGE
VILLAGE
VILLAGE
VILLAGE
VILLAGE
TOWN

Vikings could be assumed to have captured the Isle of Wight which they are using as a base for their raids. The campaign can become a land and sea affair by the British raising ships with which to fight the Vikings on their own ground, as did King Alfred.

The campaign that was chosen and is described in these pages involved a fleet of six Viking ships leaving Scandinavia (one of them, the chief's ship, being larger than the others). The fleet reaches the English coast, and splits up into individual raiding parties who move to land and capture towns and villages. At once this campaign takes on an interesting aspect because the weather has to be taken into consideration and sudden storms may well cause the complete loss or delay of one or more of the ships from arriving off the coast of Britain. Presumably the time of the year will be the raiding season of Spring or Summer and the usual methods of deciding the weather will prevail. At regular daily intervals (using the War Diary) a dice will be thrown for each ship.

1 scored will mean that the ship has been lost with all hands.

2 or 3 scored means that the ship arrives later than the rest of the fleet (the length of its lateness can be determined by dice or some other similar method).

If 4, 5 or 6 are scored on the dice then the ship is considered to have satisfactorily weathered any storms and to have arrived on time.

In the campaign under discussion two methods of Viking approach were considered. First, a dice was thrown, its score corresponding with numbered sectors on the map so that the ship (presumably through weather or tide) landed at a point on the coast not necessarily to its commander's choice. The second method was for the fleet to split up into groups or singly and go ashore at points of their choice to attack villages. A further possible complication could be the existence of a certain amount of jealousy or rivalry between the captains of the ship sso that two ships might choose the same apparently rich town or village and fight between themselves for the right to loot that area.

At any event, there was always the likelihood that the longboats might have to beach at assorted times dependent upon their point of arrival on the map and whether or not they were delayed on their journey. The overall commander of the fleet can either arrange beforehand for all ships to hold off until the fleet is in position or else the ships act independently and beach in their own time. The latter course seems more natural as a commander and crew greedy for loot will be only too keen to take advantage of an early start that allows them to move on to another village before their rivals have landed.

At the time of this campaign, the British had become accustomed to these seasonal raids and had taken steps to defend themselves. All the able-bodied men in a district were bound to band together as a levy or Fyrd to go to the rescue of an attacked area and attempt to drive off the invaders. This means that an attempt will always be made by the inhabitants of an attacked village to send a messenger to the nearest town at which the Fyrd are assembling. Knowing that it is the raiding season, it can be assumed that the Fyrd have begun to collect at the town and will be able to move off at the specified rates of map movement as soon as a messenger arrives to acquaint them with the area in which the operations are taking place. It may well be that because of the difference in landing times of the invaders that the first village to be attacked will get their messenger away so that help comes before the full strength of the six Viking ships has spilled ashore.

These messengers will be subject to the normal vagaries of 'courier cards' (see chapter 19, *Advanced War Games*). In addition facilities can be built into the rules to allow the Vikings to attempt to stop the couriers by chasing them on stolen horses or by catching the village so much by surprise that there is no time for a messenger to leave.

The campaign map shows the shore and beaches; villages are marked and an inland town at which the Fyrd are assembling. The mouths of rivers dot the coastline and the winding course of the waterways spread inland over the map. It is a part of England known intimately to the villagers who have lived there all their lives—it is less well known to the Viking invaders than they imagine.

Jumping more than a thousand years ahead, it has been said that General Slim was convinced that the British Army fought all its battles in the pouring rain on the side of the hill and at a point where two maps join each other. During this campaign in the 9th century, it may well rain and battles might be fought on the sides of hills—maps play a similarly important part. Each Viking captain has his own map of this stretch of coastline, carefully drawn and compiled from personal experience, from information given him by others who have gone before and from what he could 'crib' from the maps of rival captains.

In our campaign this factor has been covered by giving each Viking ship a map that is slightly different to the maps held by their fellows in the other five ships. That is to say that whilst every one of the six maps resembles the countryside over which the raids are to take place, there are certain errors and omissions such as a river becoming shallow before the map indicates, or a village being perhaps a mile away from its marked location or protected by sheer unclimbable cliffs unmarked on the map. Jealous of their maps and eager to gain every possible advantage over their comrades and rivals, each Viking captain permits no one but himself and his crew to see his map. None of the five captains and their commander are aware which of their maps are correct; in fact they have no reason to doubt their accuracy. Of course the umpire controlling the game has a completely accurate master-map and it is assumed that the villagers and the Fyrd are familiar with the ground and will make no errors of location.

The question of weather has already been discussed and it has been indicated that the time of the year and the weather for the opening day of the campaign must be decided and so marked in the War Diary. Steps must be taken to discover if and when the weather changes and the immediate effects of these changes are marked in the War Diary, subsequently to affect operations on the table-top. It is most important that the direction of the wind should be decided at the start of each day's map-moving and checked at prearranged intervals throughout the day to see whether or not it has changed direction. As any reader who 'messes around with boats' knows, life is always a little more difficult in sailing vessels when the 'wind is blowing against the tide'. If a Viking ship is attempting to come in on the tide and a strong wind is blowing from the shore then it is going to make it far more difficult to beach that boat than would normally be the case. This is an essential factor that will need to be decided either on the map or at some stage during the war game.

In addition to discovering the direction of the wind at the start of the first day of the campaign it is important to remember to find out how the tides are running. For example, let us assume that it is high tide at 8 a.m.; the tide remains up for one hour so that at 9 a.m. the water commences to recede. It takes three hours for the tide to go out so that at 12 noon it is low tide —a situation that prevails for one hour. At 1 p.m. the tide begins to come in and by four o'clock it is again high tide and at 5 p.m. the tide commences to go out again. With this knowledge it is not difficult to work out a tide table. Obviously, the captain of a Viking longship is an experienced sailor, likely to take every advantage of wind and tide so that he will only put himself in the position of having to beach his ship under disadvantageous conditions if there is no other possible course. This factor of delayed or difficult beaching of the longship has an important bearing on the battle because it makes it more likely that the villagers will have prior warning of the arrival of the invaders and will have time to assemble.

We have already mentioned the fact that the Fyrd automatically begin to assemble as soon as the raiding season approaches. This means that they may be ready to march but it will have to be decided how many have gathered and, in the event of an incomplete muster, whether those who have arrived march off whilst the late arrivals follow in an attempt to catch up. If a Viking crew is attacked by the Fyrd (a factor not to their liking because the Fyrd is a better organised and trained group of fighting men than the inhabitants of the coastal villages), then the Vikings can send couriers to their comrades at other parts of the area for assistance. The Viking messengers can go on stolen horses and travel in accordance with the usual conditions governing the use of such couriers. Aid can come by sea, by river or by rolling the ships overland on logs, by foot or on stolen horses. On the other hand, jealousy or greed may stand in the way so that one Viking party will not aid the other if there is a chance of loot in the other direction. All these factors can be decided by dice throws, drawing lots or by the decision of an umpire.

It has been mentioned that the Vikings may sail their boats up rivers—it was said that where there was a foot of water then a longship could go. At the same time the Vikings had a very disturbing habit of manhandling their vessels out of the water and moving them overland by rolling on logs placed under the keel. Movement rates must be specified at the start of the campaign for such methods by which it may be possible for the Vikings to outflank or out-manœuvre their enemy.

Now we come to the most essential factor of the whole campaign, one upon which it can stand or fall—that of morale. Obviously after years of devastating attacks by these incredibly fierce men from the north the English inhabitants of the coastal towns and villages, probably peaceful men of the soil, have got a thoroughly bad inferiority complex. On the other hand it may be that the Vikings, through the sheer continuity of success,

H

may lull themselves into a false and dangerous state of contempt for their English adversaries. All these factors have to be realistically reflected if the campaign is to proceed along lines that are both enjoyable and that accurately simulate the events of a thousand years ago.

One essential basic aspect upon which the morale situation must be built is that of the powers of command, bravery and tactical knowledge displayed by the commanders of both the Viking and the English forces. So far as the Vikings are concerned, the leader of the fleet should be an 'Exceptional' commander. To balance that, it is reasonable to allow that one ship has a poor crew commanded by a 'Below Average' leader. To summarise—there are six ships, one of which is a larger flagship commanded by an Exceptional commander; there is also one other longship commanded by a similarly Exceptional commander whilst three of the boats have Average commanders and one is a Poor commander.

So far as the English are concerned a dice should be thrown for the leaders of each village—a score of 5 or 6 means an Exceptional commander; score of 3 or 4 means an Average commander; score of 1 or 2 means a 'Poor' commander. The Fyrd commander would presumably be a leader chosen by his fellows for some outstanding ability or qualities so that it is not reasonable to assess him as a Poor commander so dice for his grading—1, 2 or 3 means that he is an Average commander, 4, 5 or 6 means that he is an Exceptional commander. But his two subordinates must be diced for in the same way as the village leaders.

At the commencement of the campaign, the Viking morale will be considered to be excellent and will not be questioned until some adverse situation in conjunction with the grading of a leader might cause it to fall. The presence of a Berserker will have the effect of raising morale in his immediate vicinity—some sort of ruling can be made to the effect that all men of his own side within a 6 in. radius of a Berserker will go up one step in their immediate morale.

So far as the morale of the English is concerned, there are three questions that have to be answered:

1. Do they fight, run or surrender?
2. If they fight, when and how is their morale tested?
3. Does English morale rise or fall with defeat?

It is reasonable to assume that the English, with the knowledge of the heavy price they have to pay for failure or defeat, will have some sort of factor that reflects desperation—the will to resist beyond their normal capabilities because they know that the only alternative is death for themselves and their families. As it would not be an unlikely thing for them to fight desperately it can reasonably be ruled that a dice throw of 5 or 6 will give that added factor to any group of men or individual for whom such a decision has to be made.

The first morale situation likely to arise in this campaign is when the English first become aware that the Vikings are upon them—then they have to decide whether to fight, to flee or to surrender. The answer to these three questions is dependent upon:

(a) Whether the English are surprised or not.
(b) The respective numerical strengths of the English and the Vikings.
(c) The rating of the English chief.
(d) Whether or not they have a desperation factor.

To convert this into practical terms let it be assumed that the following additions and subtractions are made from a single dice throw:

Minus 3—the English have been surprised.
Minus 1—although warned, the English have little time to prepare, i.e. Vikings loom out of the dawn mist.
Plus 1  —the English have been warned and have time to prepare, i.e. the ship is sighted off-shore or the Vikings land in daylight.
Minus 2—the English and the Vikings have approximately the same numerical strength.
Minus 1—the English outnumber the Vikings 3–2.
Nil      —the English outnumber the Vikings 2–1.
Plus 1  —the English outnumber the Vikings 3–1.
Plus 2  —the English leader's rating is Exceptional.
Plus 1  —the English leader's rating is Average.
Minus 1—the English leader's rating is Poor.

Therefore to decide whether the English fight, flee or surrender—throw a single dice, make the appropriate additions and subtractions and then act in accordance with the following table.

Dice score of 5 or 6—The English fight.
Dice score of 3 or 4—The English flee.
Dice score of 1 or 2—The English surrender.

*Examples*

An English force is surprised—minus 3.

The English outnumber the Vikings 3–2—minus 1.

The English leader is Average—plus 1.

The English have a desperation factor—plus 2.

Equals a total of minus 1.

The dice throw is 6 which less the minus 1 comes to 5 which means the English fight.

The Vikings attempt to land in daylight so the English have time to prepare gives them a plus 1.

The English outnumber the Vikings 3–1—plus 1.

The English leader is Poor—minus 1.

There is no desperation factor—total plus 1. The dice is thrown and comes up 3 which gives a total of 4 which means that the English flee.

During the actual fighting occasions may arise when it is necessary to test the morale of one or both sides. In the case of the English their morale will be tested when:

(a)  A leader is killed.
(b)  A quarter of their total strength is killed in a single move.
(c)  When a Berserker attacks.

When assessing this morale, it is suggested that such factors are considered as the respective numerical strengths; the leader's rating; the arms and equipment of the respective forces (i.e. whether a man is equipped with shield or armour, etc.), plus (a), (b) and (c) above.

There are various methods of handling the morale of both sides throughout the campaign. For example, the campaign can start with the Viking morale at a high level and the English morale correspondingly low and remaining so throughout the campaign with the exception of the Fyrd whose morale would probably be higher because they found consolation in their superior numbers and organisation. An interesting addition can be made to this campaign when an umpire is available. Probably through anger and desperation the morale of the English has risen since the previous year's raids to a level higher than the norm known to the Vikings. The degree at which their morale has risen, the point at which it is disclosed is revealed by the umpire only to the English so that the Vikings remain in ignorance of the hardening of morale until the time comes for it to be tested. A third but unexplored approach lies in a method which will tie-in the lowering morale of the English with a corresponding-ly fluctuating Viking 'careless or arrogance' factor.

*Sequence of the Campaign*

When the time arrives for a table-top terrain to be set up, then it is erected in strict accordance with the *English* map held by the umpire. Ideally the umpire should be responsible for setting the terrain because this will realistically deny the English the advance knowledge from what particular part of their coastline to anticipate danger. Should the Vikings be attempting to land in darkness or at daybreak, then the following methods of operation is suggested.

Using their not-necessarily accurate map, the Vikings estimate the distance from the edge of the map to the shore and inform the umpire accordingly. At the same time they tell him their estimation of the state of the tide, i.e., is it fully in, half out or fully out? They already know the wind direction. The umpire calculates, using the English (accurate) map whether the Vikings will and as and when they expect, taking all these factors into consideration. At the very beginning of operations, a weather card will decide whether it is misty or foggy—other factors which will have a marked bearing on whether the Vikings land safely, whether they land undetected or whether they decide risk landing at all. Rules can be made for the Viking ship to ground, strike upon rocks or otherwise be detrimentally affected by a heavy sea or by tide running against the wind.

The English will have a sentry, probably positioned on a convenient headland, whose job will be to sight invaders and sound the alarm. This man is a villager, probably a farmer or a fisherman and not a soldier. Lacking the sense of vigilance that military discipline might give a man so that he is fully alert after many cold and weary hours of staring out into the black void that is the sea, this sentry might well be half asleep. The alertness of the sentry can be determined by deciding whether he is a good, average or bad watcher—a dice is thrown and a score of 1 or 2 means that he is a bad sentry and that he will not sight the invaders until they are almost upon him.

A score of 3 or 4 means that he is an average sentry and will spot the raiders at the moment of landing.

A score of 5 or 6 means that he is a good sentry and that he will spot the raiders as they come into shore.

In each case the sentry will, at a stage compatible with his rating as a sentry, sound the alarm and fire the beacon. Until this is done, no movement is permitted among the English villagers who are considered to be asleep in their huts.

Concealment, as all experienced war-gamers can testify, is the most difficult and perhaps almost impossible factor to represent on a war-games table. So far we have been dealing with movements on the map with the campaign not yet transferred to the table-top but the time has come when markings on a map must be converted to the movement of figures over a realistic shore and sea terrain. How, except in the imagination, can we simulate the landing of the Viking raiders in darkness or the pale light of dawn? One simple method is for all the table-top operations at this stage to take place by the light of a single very low-wattage blue lamp above the table. Movement of troops is difficult for both sides and this method has the merit of irritatingly representing the frustration of attempting to move or to judge the movement of others in a very dim light.

Another practical method is to construct 'Concealment boxes' consisting simply of those long narrow cardboard boxes in which chocolate mint sweets are packed. The author uses two types of concealment box, both of them are sprayed with matt black paint, a factor which lends them a peculiarly appropriate appearance. In use, a concealment box is placed with its base facing towards the enemy and its open end back towards its user, inside the box stand the figures intended to be concealed (as these are probably on movement trays it is not difficult for them to stand rather than to unrealistically lay in the boxes). The boxes are moved at whatever appropriate rate has been decided for manœuvres in darkness or semi-darkness and the figures within are not revealed until the boxes come within a short pre-arranged distance of the English.

The second type of concealment box can be used when the Vikings approach nearer or if they are moving in a mist that might well give them partial concealment but allow their moving forms to assume a frighteningly vague and unearthly aspect. This second type of box has the bottom removed to be replaced by an appropriately cut piece of thick semi-transparent plastic. When the figures are placed in these boxes and the lid of the box put on so that there is no light coming in at the back, it is just possible to detect that there are figures within but details of their type and number are almost impossible to ascertain. It might well be that concealment boxes will be deemed suitable for use by both sides because, after all, the Vikings have no prerogative of night or the dim light of dawn.

In the event of a landing by day or should the Vikings be detected before landing, then there will be a moment when they are perhaps more vulnerable than at any other time. As their boat beaches, for a split second they will be thrown into some sort of confusion as they down oars and seize their weapons before flinging themselves over the sides of the longships, into the surf and up the beach. It may be that their estimation of distance has been wrong so that their longship grinds shudderingly to a halt or that the rowers are thrown into some sort of puzzled confusion because the shore appears to be much further away than they expected. In either event, factors should be built into the rules allowing the English (should they have overcome the Viking surprise factor) to take advantage of this momentary semi-defenceless period in the invaders' approach. Returning to concealment boxes, perhaps the simplest rules for their use is to decide that if Vikings are seen or suspected then they will land and move up the beach in a concealment box; if they are not seen or suspected then they are placed in position on the war games table as though they land and had come up the beach without being seen.

Concealment boxes may also be used if smoke from a burning house or village masks the movements of either or both Vikings and villagers. In this case the direction of the wind has a marked bearing on what follows and the umpire will be the arbiter as to how and when men are revealed emerging from the friendly shelter of smoke drifting on the wind.

Although the arms of both Vikings and villagers might well resemble each other, it is reasonable to assume that those of the invaders will be of a better quality, in better condition and certainly wielded by more experienced arms. Both sides have their own differing style of fighting. The English endeavour to form up in the traditional shield-wall where each man is protected by his shield in front of him and can act as a guardian to the man on his right or left who is similarly performing on his behalf. Should the Vikings catch the English before they are able to form this wall then the English are at a great disadvantage but if the shield-wall is formed then the rules should give the English greater parity with their fiercer opponents. The Vikings fight individually and not in a shield-wall—they had a way of fighting back-to-back which should give them increased powers. It has already been mentioned that under certain conditions a Viking can go berserk with a markedly fearful effect upon his opponents and an encouraging reaction to those of his own side. It should be decided beforehand how many men are likely to be given the opportunity of achieving this ferocious state—say one man in ten is selected, that man being specifically designated and, at any given moment in the mêlée, a dice may be thrown to see whether or not he goes berserk. This state should not be achieved too easily as it might seem that berserkers were not as prevalent as the chronicles would have us believe. In addition, it must have been an extremely wearing reaction and invariably the man, bereft of all armour and protecttion, finished up dead. It is suggested that to bestow upon a selected figure the right to go berserk two dice should be thrown and a total of 11 or 12 is required.

Neither English nor Vikings fought on horseback but both used horses to move more quickly from place to place. The English, particularly the Fyrd, probably had horses tethered nearby and could use them to make their escape when beaten. It was a known Viking practice to steal horses from a defeated village and thus move more quickly on to the next selected victims. This must have been an interesting sight—rather like a gang of drunken sailors on the spree! A practical note to consider is that horses minus riders should be in fields near the huts or perhaps kept in the huts themselves (there are many horses in the various Airfix kits that ideally lend themselves to this role.

This is an unusual type of war-games campaign with many facets that may be new to the reader. It can be carried on by two, three or even more war-gamers and has the merit of not requiring many troops and those that are used can be obtained quite cheaply. Miniature Figurines turn out a very nice line of Vikings whilst Woolworths sell 'Empire-made' Viking ships imported into this country by Airfix. With the ship goes a crew of about 20 or 30 very inferior plastic 20 mm Vikings. These figures do not justify very much time spent on painting but, stuck on stands in twos and with a minimum of paint work, they represent a cheap and readily available source of figures for the campaign. The ships lend themselves admirably to our purpose—they look much more realistic if the hulls are sprayed with matt black paint to represent the pitch-encrusted wooden hulls of the original longships. The English villagers can readily be converted from the Airfix 'Ancient Britons' box or the 'Robin Hood' set.

Three aspects of the battle between the French and the Austro-Bavarian armies in Italy, 1859. (Chapter 22 'A Narrative Battle')
*Above*   General view of the battlefield, from a point behind the French left/centre
*Centre*   Austrian infantry battalions attack the French right (viewed from a point behind the cemetery)
*Below*   Another view of the Austrian infantry attack with Bavarian Uhlans moving forward. Austrian Dragoons and Hussars move forward in support (viewed from the Bavarian position on the crest of the ridge)

Jackson's 'Foot Cavalry' storm a Union-held churchyard during the Manassas Gap battle. (Chapter 23 'Stonewall Jackson in the Shenandoah Valley')

*Above*  The battle by the river at Fort Republic, where General Sheridan and his Union force destroyed Jackson's hopes in the Shenandoah Valley. (Chapter 23). This battle was 'away' from home—on a terrain set up by wargamer Peter Gilder—note the vast difference in appearance between this and the sandtable terrain used in the other battles of this campaign

*Below*  General Getty's Federal force being attacked by Jackson's larger Confederate army in the Manassas Gap. (Chapter 23)

A later stage of the Confederate attack on the smaller Union force
at the Manassas Gap—Getty's sorely pressed men are being
slowly forced back, but nightfall is coming to their rescue.
(Chapter 23)

# 15

# The Agincourt Campaign

One of the most fiercely patriotic periods in English history must surely be the Hundred Years War, when England began to emerge as a nation to be feared and respected. It was the last era of land warfare before gunpowder enabled the gunner and the musketeer to set his irrevocable stamp upon the strategy and tactics of warfare. Changing warfare from a brawl between armed mobs, Edward I of England and the Kings that followed him had skilfully combined the knight and man-at-arms with the incomparable English archer whose longbow was probably the most devastating weapon known to mankind until the advent of the machine-gun. Blooding his fighting methods and weapons in wars against the Scots, Edward I took on the numerically stronger and arrogant French. The battles of Crecy, Poitiers and Agincourt, together with a host of lesser-known affairs, laid the foundations of English military supremacy that was to last for 600 years.

Known as the 'Agincourt Game' the short campaign that follows is won or lost in the knowledge that the French have to get to grips to be victorious whilst the British have to hold them off or be destroyed.

In a war game permitting considerable tactical manœuvring, the small British force under Henry V are moving across country in an effort to get across the River Somme before being cut off by the much larger French pursuing force. At the mouth of the Somme lays the English fleet waiting, in a Dunkirk-like manner, to pick up and carry to safety the English army that has been wending its ravaging way across France.

The map is so drawn that the English, in spite of their faster rates of movement, are hard pushed to reach the safety of their boats without turning and giving battle to the pursuing French. The French do not have to attack the English but if they do not bring them to bay then the smaller English force will get away so it is definitely in the French interest to force a battle. Anyway, if they don't then there will not be a campaign! Certain points on the map provide very advantageous defensive positions for the English to make a stand. To give battle in the open, particularly on the extensive plain that stretches inland from the port, is to invite disaster because of the large numbers of mounted knights and men-at-arms who form a great percentage of the French force.

Both forces come on to the map at the north-western corner (marked with an X)—the English having a start of one game-move. The map is scaled at 1 inch to the foot and is formed of four war-games tables.

At the bottom righthand corner of the map, the two river mouths are marked A and B. At low tide it is possible to ford both these points. It has to be decided (and shown on War Diary) the state of the tides so that a force may either cross at once or be delayed. An ideal example of such an occurrence was the English crossing of the ford at Blanchetaque during the journey to Crecy in 1346. (See pp. 97–9, *Bowmen of England* by Donald Featherstone.)

The rates of movement on the map are given below and it is suggested that these move-distances are scaled to use on the war-games table, so that dismounted men-at-arms for example will move across the plain at the rate of 1 in. per map-move or 12 in. on the war-games table.

| | Cross Country | Plain | Marsh | Beach |
|---|---|---|---|---|
| Archers | $1\frac{1}{4}$ in. | $1\frac{1}{4}$ in. | $\frac{1}{2}$ in. | $\frac{3}{4}$ in. |
| Dismounted Men-at-Arms | $\frac{3}{4}$ in. | 1 in. | $\frac{1}{4}$ in. | $\frac{1}{2}$ in. |
| Crossbowmen | $\frac{3}{4}$ in. | 1 in. | $\frac{1}{4}$ in. | $\frac{1}{2}$ in. |
| Light Horsemen | $1\frac{1}{4}$ in. | $1\frac{1}{2}$ in. | | |
| Knights and Men-at-Arms | $\frac{1}{2}$ in. | $1\frac{1}{4}$ in. | | |
| Waggons | $\frac{1}{2}$ in. | 1 in. | | |

The light horsemen, knights and the waggons cannot move on the marsh or the beach.

At bridges, fords, forest defiles and mountainous tracks, it is considered that there will be an inevitable 'piling up' of the columns so that speed is cut by a half at these points.

The French have a train of waggons carrying their food—under no circumstances will these waggons be abandoned. On the other hand, the English force are living off the land and the few waggons they possess will carry food when available and the armour of their knights or men-at-arms. These waggons may be jettisoned if required, in which case the knights and

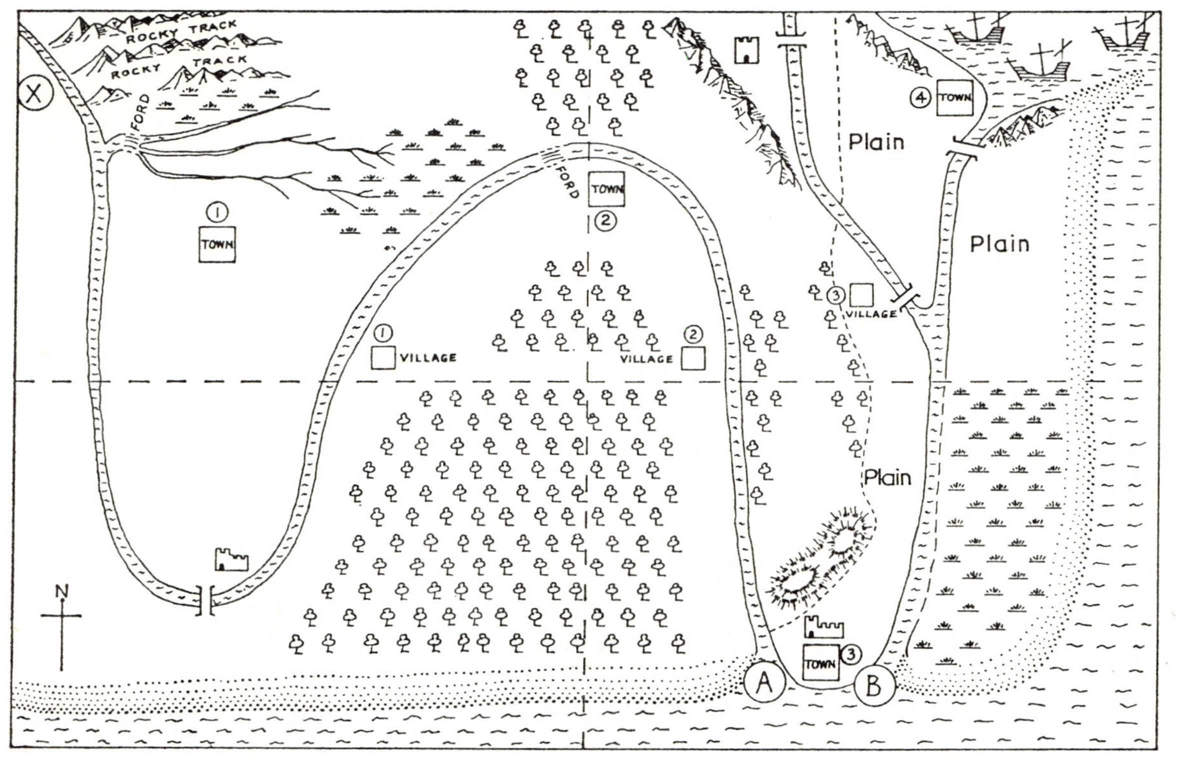

ROCKY TRACK
ROCKY TRACK
FORD
FORD
X
TOWN
TOWN
VILLAGE
VILLAGE
VILLAGE
TOWN
TOWN
Plain
Plain
Plain
N
A
B

men-at-arms will be required to march in their armour which will cause the wearers to move at half their normal speed. To secure their food, the English must loot the villages and towns they encounter on their route. This means that either the food from the villages has to be brought back to the main body of the army in waggons or else the main body has to go to the food. The method of simulating these factors are as follows:

A village equals 1 unit of food and a town equals 3 units of food.

1 unit of food keeps the English army moving at normal speed for one day or at half speed for two days.

To loot a town, it is necessary for an English force to be 100 points strong whilst to loot a village requires a force of 50 points strength. These points values are made up by counting mounted knights and men-at-arms as 3 points.

Dismounted men-at-arms (with shields) count 2 points.

Dismounted men-at-arms without shields count $1\frac{1}{2}$ points.

Archers and crossbowmen equal 1 point each.

If a town or village has been reinforced by French troops moving ahead of their main body, then it might be necessary for a small battle to be fought. In this case the inhabitants of the towns and villages will probably take a hand in repelling the English invaders. For this purpose it can be assumed that each town has 60 inhabitants whilst each village has 20 people living within it. These villagers are peasants of the lowest order, lacking adequate weapons and likely to be in awe of organised soldiery. So they must have a lower points value and a much lower morale rating than the soldiers with whom they might be allied or against whom they might fight. The situation could be handled in a manner similar to that of the Viking campaign, when invaders land and the English inhabitants of the coastal villages have to decide whether they are going to fight, flee or surrender. It is possible that any war game resulting from such an encounter would be a small affair unlikely to justify setting up a fairly involved terrain or allowing such an encounter to take up valuable war-gaming time. It can be settled on paper by the following method:

French and British soldiers will be counted in points as already nominated whilst the villagers count $\frac{1}{4}$ point each.

When the defenders are half the strength of the attackers (counting in points) then there will be a delay of *one move*.

When the defenders are three-quarters strength of the attackers (counted in points) then there will be a delay of *two moves*.

When the defenders are equal in strength to the attackers (counting in points) then there will be a delay of *three moves*.

When the defenders are 25% stronger in points value than the attackers then there will be a delay of *four moves*.

When the defenders are 50% stronger in points value than the attackers then the latter will not attack the town or village.

Once a town or village has been taken, each attacker can bring away enough food for himself and two others. If waggons are taken along and enter the town or village then three waggons can carry away enough food for the entire English army.

The towns and small castles dotted about the map each have garrisons of 20 men-at-arms. They may be mounted if they sally out to attack the English, who have no means of besieging the towns and castles. It amounts to a situation where the English will not interfere with the garrisons providing the garrisons do not interfere with them.

At the beginning of the campaign French Light Horsemen or Foot in any desired numbers up to 50% of the entire French strength may begin the game on the opposite side of the river in an attempt to head off the English. They may bar the fords by planting stakes which will take them 3 moves to plant—it will take the British 2 moves to remove and clear the passage through the ford.

It is not the policy of this book to suggest or dictate the rules under which these battles or campaigns should be fought. Nevertheless it is considered to be worthy of mention that this campaign was most satisfactorily fought using the following set of simple rules.

*Moving*
Archers—15 in.
Dismounted men-at-arms—12 in.
Mounted men-at-arms and knights—*Walk*, 9 in.; *Trot*, 12 in.; *Gallop*, 18 in.
(After a 'Gallop-move' a 'Walk-move' *must* follow.)
Waggons—12 in.
3 in. extra per move on roads for all arms.

*Firing*
Archers fire *twice* if they do *not* move.
Archers fire *once* if they move 7½ in. or less.
Crossbowmen fire or move but not both.

| *Saving throws* | Range | 18″ | 12″ | 6″ |
|---|---|---|---|---|
| | | $\frac{1}{4}$ score | $\frac{1}{2}$ score | $\frac{3}{4}$ score |
| Mounted knights and men-at-arms (+1 for shield) | | 3, 4, 5 & 6 | 4, 5, 6 | 5, 6 |
| Archers, Crossbowmen | | 4, 5, 6 | 5, 6 | 6 |

*Mêlées. Points value*
Archers—1.
Crossbowmen—1.
Dismounted men-at-arms—1$\frac{1}{2}$.
Dismounted men-at-arms with shields—2.
Mounted men-at-arms and knights—3.
Impetus—take out 1 man per horseman (saving throws 5 or 6).
Axe and Spearmen add $\frac{1}{2}$ point each for first round of mêlée.
Commanders fight individually—add 1 per dice.
Kings fight individually—add 2 per dice.
Fight 1 dice per 5 points.
Fight in groups of approximately 20 points.
Add 1 per dice to a group with a Commander.
Add 2 per dice to a group with the King.


*Saving throws after mêlées*
Horsemen—4, 5, 6.
Infantry—5, 6.
Crossbowmen, Archers—6.

*Charging Home* when attacked by cavalry—Infantry need 4, 5,
    6 to stand (add 1 to dice if infantry on higher ground).
Spearman and Axeman *always* stand.
If infantry stand—cavalry refuse if they throw 1 or 2 (1, 2 or 3
    if infantry on higher ground).

*Morale* is tested when:
a leader of a group is killed
a group loses $\frac{1}{4}$ in one move
when a group are below $\frac{1}{2}$ original strength at end of mêlée by
    side losing most men.
Add 1 to *all* English morale dice except when outnumbered (in
    groups) by more than two-to-one.
Otherwise, morale rules to personal choice.

# The Fight for Alton Church
### (English Civil War)

Table-top battles are always enjoyable but a much greater degree of interest is given to them if the actual war game is preceded by a build-up that brings realism and continuity. This ties in with the well known fact that a war-gamer frequently thirsts to campaign in any particular period of military history he happens to be reading at the time. This is one of the fascinations of the hobby, causing it to be both endless and boundless.

On a typical war-games evening, the table is laid with an assortment of terrain and two armies are selected. Your opponent arrives and takes one of these armies whilst you take the other. Both forces assemble on opposite baselines and the battle commences until such time as one side are the victors or the passing of time calls a necessary halt. How much better and more interesting it can be if the visitor is greeted with a neat little narrative (either on paper or by word of mouth) describing the events which led up to the forthcoming battle and allowing both combatants to attempt to out-manœuvre each other even before a single figure has moved on to the table-top battlefield.

One such 'potted' campaign began when the author read the following half column in the local Hampshire County magazine:

'Colonel Boles held the church at Alton for the King against Cromwell's troops in 1643, and it was when he was standing in the Jacobean pulpit that the Colonel died for Charles Stuart. The bullet marks on the pillars and the doorways and the door are evidence of the bitter fighting which took place at Alton.

'On the night of December 12–13th the Parliamentary forces under Sir William Waller took up position on the hills above Alton to the west. On the following morning they were discovered in considerable strength by the Cavaliers. Lord Crawford, who was in command of the Royalist troops, went along to the Winchester road to warn Hopton, a Royalist leader, and bring relief to the handful of Cavaliers left behind under the command of Colonel Boles.

'The latter, although greatly outnumbered, withdrew his handful of men to a spot on rising ground near the parish church of St Lawrence. Here they held out for some time aided by the burning of a house, the smoke of which drove in the faces of the attacking Parliamentarians. Eventually the latter rushed the churchyard, and Colonel Boles and his men were ultimately driven from the building and a fierce fight took place. The Cavaliers, holding the churchyard wall, had not retired when fighting began and most of these were slain around the door of the church. Then the door was swung to and fastened and the attacking force threw hand grenades against it and through the windows and after a stout resistance the door crashed inwards, burst from its hinges, and the Parliamentarians swarmed into the church.

'The Cavaliers had piled their dead horses in the aisles and behind these they fought stubbornly, but in the end they were overwhelmed by numbers, killed almost to a man and their gallant commander, Richard Boles, fell, fighting to the last.

'The inscription on the pillar of the nave recording his gallant deeds is the fascimile of that erected in Winchester Cathedral. Of Colonel Boles the King said "Bring me a mourning scarf, I have lost one of the best commanders in the Kingdom".'

Inspired by these words, a pilgrimage was made to St Lawrence Church in Alton where still can be seen the musket-ball holes made in the old blackened timbers of the church door. Before the homeward journey was half done, mental plans had been completed for a war game that would simulate Colonel Boles' gallant last fight of 327 years ago.

*Narrative*

In December 1643, Hopton the Cavalier commander was faced in Hampshire and on the Sussex borders by an army under the command of Waller. Seeing that there was no immediate chance of engaging and destroying Waller's army, which had withdrawn into Farnham and its castle, Hopton turned south and captured the town and castle of Arundel. In so doing, Hopton overstretched himself and a reinforced Waller pounced on a Cavalier

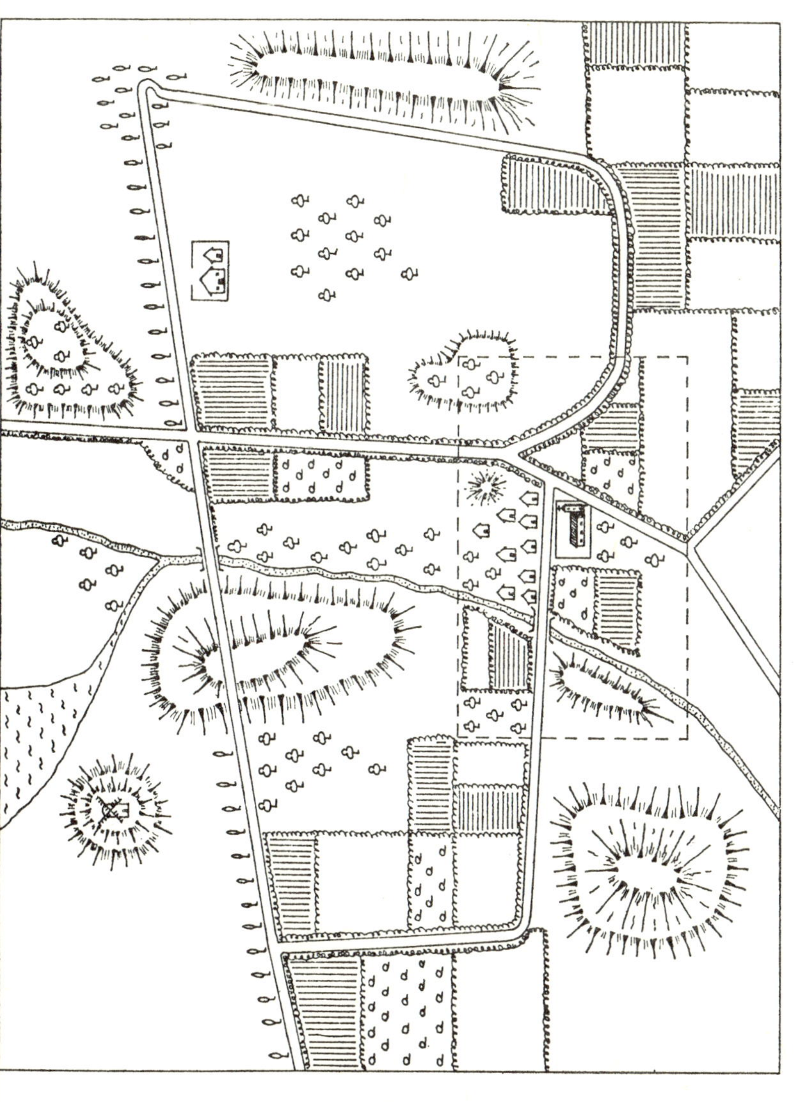

force under Lord Crawford in the vicinity of Alton. Most of the Cavaliers got away but a small part of the force under Colonel Boles became detached and were pursued by the Roundheads towards Alton, a small town on their route towards safety at Petersfield. It is at this point that our campaign begins.

*Technical notes*

The first requirement is a large-scale Campaign map. No doubt it will be possible to discover in the archives of the pleasant Hampshire town of Alton a map of the period showing roads, tracks and the layout of houses and streets. From this can be drawn a large-scale semi-diagrammatic map required for this potted campaign. However, this may well be a time-consuming refinement that the war-gamer will consider unnecessary—for him the map used in our campaign and illustrated in this book will be considered adequate. Although a modern Ordnance Survey map was consulted, the roads that are drawn on our map, although bearing a real relationship to the topography of the area, have been marked to indicate a number of alternative routes back to the safety of Petersfield which will require decisions to be made by both Cavalier and Roundhead Commanders. The map was drawn on a sheet of 1 in. squared graph paper, 22 by 17 squares in size. The scale of the map was 1 in. to 1 ft. on the war-games table—the whole therefore representing an area of approximately 4 by 3 war-games tables each 8 ft. by 5 ft. The table on which Alton Church and its environments was set out represented the actual stage for our war game.

The convenient scale of 1 in. to 1 ft. made possible the manner in which this campaign was conducted. In short the whole campaign was one large war game of which part was fought with figures on the war-games table on a terrain representing one of the twelve squares of the map, whilst the remainder of the campaign actually took place on the map although being fought exactly as a war game. Here is the manner in which it was conducted.

The respective forces were as follows:

*Cavaliers*
2 Regiments of Infantry.
1 Squadron of Cavalry.
1 Horse Gun.

*Parliament*
3 Regiments of Infantry.
2 Squadrons of Cavalry.
1 Horse Gun.


An infantry regiment consisted of 40 musketeers, 40 pikemen with one mounted colonel, 4 dismounted officers, 2 standards and a drummer. A cavalry squadron consisted of 26 troopers and 4 officers. The guns were small 'Galloper-guns' whose split-trail formed the shafts in which a single horse was harnessed—thus the gun was very light and of small calibre.

Using 20 (25) mm figures—the size of converted Airfix figures or those English Civil War pieces put out by Miniature Figurines—a complete formation of infantry would cover an area of 8 in. by 3 in. when placed four abreast on movement trays and formed up in column or 4 in. by 6 in. when set out ready for action. The cavalry squadron would cover an area of 10 in. by $3\frac{1}{2}$ in. in column or $7\frac{1}{2}$ in. by 5 in. when moving into action. These measurements represent the actual size of the formation on the war-games table. For the purposes of moving on the map, small shapes are required to represent these formations in order that map movements will be to scale. The small shapes can either be scaled down versions of the measurements already given—that is to say infantry can be $\frac{1}{3}$ in. by $\frac{1}{2}$ in. in column or $\frac{3}{4}$ in. by $\frac{1}{4}$ in. in formation whilst cavalry will be $\frac{5}{8}$ in. by $\frac{3}{8}$ in. in formation or $\frac{1}{4}$ in. by $\frac{3}{4}$ in. To avoid becoming unnecessarily complicated, the counters used in the author's campaign were $\frac{1}{2}$ in. by $\frac{1}{2}$ in. for infantry and $\frac{3}{4}$ in. by $\frac{3}{4}$ in. for cavalry. Each counter was numbered and a chart was made out showing the formation represented by each counter.

Both sides opened the campaign by coming on to the map at its northern border via the Farnham Road. A dice was thrown for each Cavalier counter to indicate how far it was in advance of its Roundhead pursuers. A dice score of 1 or 2 meant that it was allowed one complete move start on the map; a dice score of 3 or 4 gave it two complete moves start whilst a dice throw of 5 or 6 gave a start of three complete map moves. The rate at which movement took place on the map was scaled down from the move-distances allowed on the war-games table with allowances being made for the size of the unit or units that could move on roads or tracks at one and the same time. It was generally considered that any unit moving on a road or track had to be in column and needed to deploy onto the ground on either side of the road to go into any other formation. For information, here are details of the move-distances used in the actual campaign.

Pikemen (moving slowly because of the 18 ft. pike and armour) —12 in. on the war-games table, 1 in. on the map.

Musketeer (allowed to act collectively or in a skirmishing role)— 15 in. on the war-games table, 1¼ in. on the map.

Cavalry—18 in. on the war-games table, 1½ in. on the map. Cavalry are allowed a 'charge' move to make actual contact with the enemy of 21 in. on the war games table and 1¾ in. on the map.

Galloper-guns—18 in. on the war-games table, 1½ in. on the map. Allow 3 in. to unlimber and 3 in. to limber up.

25% of the distance is deducted when infantry cross a wall or when cavalry push their way through a hedge (hedges cannot be climbed or pushed through by infantry). 25% is added to move-distance when moving on a road.

As this is in part a war game fought on a map it is necessary to know the ranges of the various weapons that might be used. The musketeers can move loaded but must be stationary to fire. Firing takes up *half* of their move. Thus a musketeer who is loaded may:

(a) Move 15 in. without firing and remain loaded.

(b) Move 7½ in. and fire—taking the entire next move to load.

(c) Fire and then move 7½ in. unloaded—taking the entire next game-move to load.

The range of a musket was 12 in. It is advisable to only fire part of a regiment so that there is never a time when *everyone* is unloaded.

The cavalrymen each carried two pistols which could only be reloaded when the man was stationary, taking *half* a move to load each pistol. The method of cavalry attack was for the horsemen to trot or walk up to within pistol range of a formation of infantry and then try to force a breach by firing their pistols into the pikemen before charging into them. The range of a pistol was 6 in. on the war-games table, ½ in. on the map. A cavalryman has the alternative of:

(a) Moving 18 in. (21 in. if charging) and firing both pistols. He is then completely unloaded until he halts and devotes a move to reloading.

(b) He can move 9 in. and spend the other 9 in. of his move reloading his pistol. He cannot charge-move in the same move as he reloads.

The gun had a range of 24 in. on the war-games table and 2 in. on the map. A gun cannot move and load but may move already loaded. It takes a complete move to load so that a gun may fire in one move—move *or* load in the next move and if it moves and it does so unloaded it will take the next move to load.

The map-battle is carried on exactly the same as the war game in that a movement sheet is used and orders written down for units in exactly the same way as on the war-games table. It is necessary to detail on the movement order sheet 'formation' of the various units so that the question does not arise of one man claiming that an infantry unit were in a defensive 'stand of pikes' whereas his opponent charged them with cavalry on the assumption that they were in column, etc., etc. Similarly, a note must be made on the movement sheet of the loading of muskets, pistols and the guns because one of the arts of war-gaming in this period is to catch one's opponent unloaded.

The campaign proceeds on the map, with both sides moving, firing and mêlée-ing exactly as on the war-games table and with casualties being noted on the movement sheet. As and when a unit moves on the map into the marked out battle area, then that unit is transferred to the relevant point on the actual war-games table which has already been laid out in preparation. Eventually, when a reasonable proportion of units of both sides have actually arrived in the battle area, then operations are transferred to the war-games table and late arrivals (still moving on the map) come on to the table at specific points and times. From then on, the campaign proceeds as a normal war game. The casualties that have been marked on the movement sheets are totalled and removed from units before they are placed on the actual war-games table so that it may well be that an infantry regiment comes on to the table with perhaps only two-thirds of its strength remaining. It will have been noted as casualties occur during the map-moving whether those casualties are musketeers or pikemen so that the unit comes on to the table accurately represented in the light of what has already transpired.

Another point worthy of mention is to emphasise that the campaign is being fought in England in December 1643—a time of the year when inclement weather may have a marked effect upon troop movements and operations. Using the methods detailed in chapter 27 'Weather in War Games' of the book *Advanced War Games*, we find that for winter months the following types of weather can reasonably be expected:

Torrential rain, light rain, snow, blizzard, mist, fog, dull, bright, high wind, average.

This is not a campaign of any duration because by its very nature as a pursuit and bringing to bay of a smaller force it should take only a day or part of a day. Once the weather has been initially decided for that day it is unlikely that there will be any major changes so that the campaign we fought under the conditions prevailing at dawn on the day in question. A set of sixteen weather cards should be made out with three Average; two Light Rain; two Mist; two Fog; two Dull; one Torrential Rain; one Snow; one Blizzard; one Bright; one High Wind. Draw a card at the commencement of the campaign—the conditions shown on that card will prevail for the first 4 moves of the campaign (8 moves equal a day). At the beginning of the 5th move draw another card and let those conditions be applicable to the second half of the campaign as it is considered not unreasonable to assume that in December the weather could change dramatically at some stage of a day. The effects of the weather upon the subsequent operations are as laid down in chapter 27 of the book *Advanced War Games*.

There are a number of factors peculiar to this period of warfare that should be noted in connection with this campaign if it is to be fought realistically. Musketeers in the open without the support of pikemen could not stand against cavalry. If time allowed, musketeers formed into a square known as a 'stand of pikes', with their 18 ft. long spears presented forward in a hedge that could not be broken by cavalry unless their pistols made a breach at least two pikemen wide. The pikemen ran back into the shelter of this square and either stood between the pikemen firing their muskets or, if pressed closely by cavalry, they flung themselves on the ground at the feet of the pikemen to roll under the shelter of their outstretched pikes. The rules for the formation and use of a stand of pikes can be to the choice of the war-gamer, possibly resembling those rules covering squares in Napoleonic warfare. To avoid arguments, it is worth coming to an agreement at the start of the campaign that a stand of pikes can only be formed where there is space. It is not good enough for an infantry unit to be moving down a narrow road or lane, to be attacked by cavalry and then to claim that they have formed a stand of pikes because there would just not be enough space in that lane for such a formation. At the same time, whilst a stand can only form where there is space a rearguard turning at bay in a lane with hedges will be, in effect, a sort of 'stand' simply because the hedges will prevent the cavalry from getting at the flanks of the column as the hedge is considered to blunt a cavalry charge.

Another factor worth considering is that, as in all civil wars, the morale of all units was suspect. To begin with they might fight well then at the slightest turn everyone would flee. Or else a unit would run at the first shot only to quickly rally and fight on bravely. Sometimes they would rally but not move forward into the battle for a little while. Cavalry were not terribly well organised, often being formed of gentlemen-farmers and the like who regarded a cavalry charge rather like a hunt after a fox (Wellington had the same trouble with his cavalry in the Peninsula nearly 200 years later). Once cavalry are loosed in a charge-move they are out of control and will probably continue in their original direction until some inbuilt factor in the rules brings them under control again. One has only to read accounts of the activities of Prince Rupert's cavalry at Edgehill and other battles to realise just how big a factor this lack of control could be.

This is an interesting small campaign that can be fought by two, three or four players. Obviously the more players that take part in a campaign or war game the quicker it will run and this game is particularly suitable for more than two players. For example, if each of the Cavalier infantry regiments, the squadron of cavalry and the gun are in different hands without any consultation being permitted because of the haste of their flight then independent action is required from each which will have a far-reaching and realistic effect upon the battle that ends the campaign. Similarly, a dissemination of command amongst the pursuing Parliamentarians may well allow a reasonable proportion of Cavaliers to get away.

# 17

# The French and Indian War
## (Mid-18th Century)

The Seven Years War (1756 to 1763) forms a most fruitful and fascinating era for the war-gamer to study. Conflict with France, Austria, Saxony, Sweden and Russia on the one side against England, Hanover and Prussia on the other. Operations took place in India, Europe and North America and it is in the last continent that this campaign took place.

In chapter 15 of the book *Advanced War Games** 'The Composition of Force' details are given of two very large (by war-games standards) battalions of British and French infantry of the mid-18th century. Both are the 37th Regiments of their respective countries—the 37th Regiment of Foot (Hampshires) of Britain and the 37th Regiment of Foot (Royal Roussillon) of France and their composition is as follows:

1 Grenadier Company composed of 36 officers and men.
1 Light Company composed of 37 officers and men.
8 Line Infantry Companies each composed of 25 officers and

men, plus colour parties and bandsmen, etc., giving a total of about 270 men each.

Each of these men was a 30 mm plastic figure made by R. W. Spence-Smith (alas, no longer in production). All of these figures were in the marching position so that the entire regiment has a complete uniformity which appeals to the author. Companies are distinguished by having their number or letter painted on their base. Although intended to fight against each other in large-scale, rather stylised battles, there will possibly be times when it is necessary for these large regiments to be split up into a size more compatible with normal war-games requirements. Each regiment has its own colour party bearing the national and regimental flags—the Hampshires with their yellow facings bearing a colour of that background with the necessary embellishments and a national flag of the period

* *Advanced War Games* by Donald Featherstone, Stanley Paul, London, 1969.

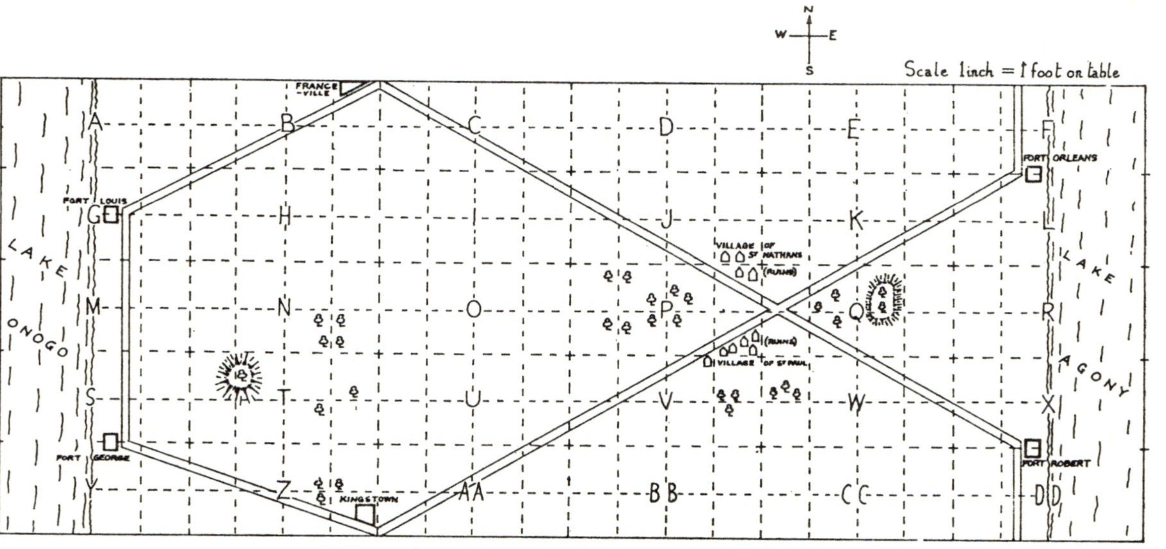

N
W E
S
Scale 1 inch = 1 foot on table
A
B
C
D
E
F
FRANCE-VILLE
FORT ORLEANS
FORT LOUIS
G
H
J
K
L
VILLAGE OF St NATHANS
(RUINS)
LAKE
ONOGO
M
N
O
P
Q
R
(RUINS)
VILLAGE OF St PAUL
LAKE
AGONY
S
T
U
V
W
X
FORT GEORGE
FORT ROBERT
Z
KINGSTOWN
AA
BB
CC
DD

whilst the French have a white standard with gold fleur-de-lys of France plus the regimental standard of the Royal Roussillon Regiment. In addition three other sets of colours for each country have been manufactured—the three British for regiments also with yellow facings and three French standards for regiments with the blue facings of the Roussillon Regiment. Thus it is possible to split the large battalion up into four regiments each of about 50 officers and men plus a proportion of the Grenadier and Light Companies, giving each a colour party so that they may take their place as four separate regiments.

The colourful activities of the Seven Years War on the vast canvas presented by North America and Canada are so colourful as to present an almost irresistible challenge to the author. Besides the formal manœuvres of the red-coated British and the white-uniformed French, there are their less civilised and disciplined allies, the Indians with the French and the Rangers with the British. The campaign that follows was only intended as a small affair, fought over a map so manufactured and balanced as to be more or less equal for both sides, giving each ample opportunity for tactical manœuvring and the necessity of making choices and decisions. The objective for each side was to break through the defended border area and into the territory behind. To do this the British would have to advance through the base-town of Franceville or take Fort Orleans and proceed along the road stretching from it; the French could only advance through the British base of Kingstown or by capturing Fort George and then proceeding down the road that ran back from that defensive post. It was not considered that either force could or would attempt an invasion across country.

The forces taking part in this campaign were not scaled down versions of larger forces but actually fought in the strengths shown below. Dispositions were as follows:

*British. 37th Regiment of Foot (Hampshires)*
1 Grenadier Company (36 officers and men).
1 Light Company (37 officers and men).
8 Line Infantry Companies (each 20–25 officers and men—
total 183 officers and men).
Roger's Rangers (50 men).
1 Troop Horse Grenadiers (12 men).
1 Squadron Dragoons (24 men).
1 Gun.

*French. 37th Regiment of Foot (Royal Roussillon)*
1 Grenadier Company (37 officers and men).
8 Line Infantry Companies (each composed of 25 officers and men—total 183 officers and men).
Indians (120 men).
1 Squadron Lifeguards (24 men).
1 Troop Burgundy Gendarmes (12 men).
1 Gun.

Both bases—Kingstown and Franceville—have their own garrisons. At Kingstown the British have the 1st Regiment of Foot (64 officers and men) whilst Franceville is held by the Maison du Roi Regiment (60 officers and men). In addition, at the onset of the campaign the British must garrison Fort George and Fort Robert and any other places they desire may be similarly occupied within the British half of the map. In addition to Franceville, the French must garrison Fort Louis and Fort Orleans together with any other places within the French half of the map.

Both sides are endeavouring to force the frontier between the lakes. The Grenadiers, the Line Infantry, guns and cavalry can only move on the roads whilst the Indians, the Rangers and the British Light Company can move anywhere. Both sides have six canoes each of which holds 10 men together with two boats carrying 20 men each which are able to tow three rafts each carrying 16 men. For the purposes of transportation one gun equals 8 men.

Both sides have a guide but he can only lead one party at a time. This party will be aware of the ground that lies ahead of them but any other force without a guide (except Indians and Rangers) will not know what lies ahead. In the campaign these factors were simulated by erecting a curtain across a blank war-games table and the French laying the terrain on their side of the curtain and the British laying the terrain on their side. Then initial troop dispositions were made and the curtain lifted.

Before any actual operations could take place on the table-top battlefield it was necessary to conduct a considerable amount of tactical manœuvring on the map. This was carried out at the following rates:

Grenadiers, Line Infantry and Guns—1 square per game-move.
Cavalry—1½ squares per game-move. (These forces can only move by road.)

Indians, Rangers, British Light Company—1½ squares per
  game-move.
Canoes—1½ squares on the lake per game-move.
Boats—1 square on the lake per game-move.
Rafts (towed by boats)—1 square on lake per game-move.

(Same speed as boats at all times.)

There are two convenient methods of carrying out map-moving in a campaign of this nature—by the 'matchbox method' and by each side marking up tracings of movements on a transparent plastic map-cover, an independent umpire comparing these tracings to decide whether or not contacts have been made.

As all the six war-gamers involved in this game wished to actually fight and not to umpire, the matchbox method of moving was adopted. The three French commanders took up their positions in the dining-room whilst their three British counterparts made themselves comfortable in the lounge, the essential matchbox-device being transported back and forth by the author's children who readily volunteered to act as runners. Being one of the British commanders, the author can testify to their plan of campaign which, at the time, seemed to hold the merits of being bold, unorthodox and unexpected. Briefly it consisted of leaving two Line Companies at Fort George on the shores of Lake Onogo on the British left flank; Kingstown was garrisoned by the 1st Regiment of Foot; the village of St Paul's was occupied by a Line Company with cavalry close at hand; the Rangers were lying to the right of St Paul's with orders to circle right to take St Nathan's from the rear; on the left side of the village of St Paul's the Light Company lay waiting to circle left and, in combination with the Rangers, to assault St Nathan's in the rear. The Grenadier Company and five Line Companies with the gun were in and around Fort Robert by Lake Agony on the British right flank. The plan was for St Nathan's to be taken and held until the dispositions and intentions of the French became apparent. If the French were in some strength at Fort Orleans with the intention of either taking Fort Robert or moving down the road to take Kingstown, then the British would attempt to hold Fort Robert and the villages whilst pushing a strong force with cavalry up the road to Franceville (the guide would go with this party). If the French were in strength between St Nathan's and Franceville then operations would be directed at Fort Orleans by an approach from the lake. At that time the plans of the French were not known.

Eager fingers placed numbered counters into their respective matchboxes and, on the very first move, a contact was made at the crossroads between the ruined villages of St Nathan's and St Paul's. With the crossroads as the centre point, a terrain was set up on the war-games table that covered the area marked on the map. The command posts in the dining-room and the lounge were evacuated and the opposing generals trooped upstairs to the war-games room.

When both sides had laid down their forces it could be seen that the French had garrisoned the ruined houses of St Nathan's with the Grenadier Company and one Line Company of the Royal Rousillon Regiment with a force of about 20 Indians in the woods to the left rear of the village. The British Line Infantry Company occupied the ruined houses of St Paul's with Roger's Rangers spread out on the right of the village and the Light Company on the left. It will be remembered that these two groups had been ordered to circle wide of St Nathan's on the right and left respectively to take it in rear. The Light Company on the left was supported by the squadron of Dragoons whilst the Horse Grenadiers lay in reserve on the right of the village behind the Rangers. Numerically, the French were well out-numbered—the actual forces consisted of:

*French*
37 Officers and Grenadiers.
25 Officers and Line Infantry.
20 Indians.

*British*
25 Officers and Line Infantry.
50 Roger's Rangers.
37 Officers and men of the Light Company.
24 Dragoons.
12 Horse Grenadiers.

In spite of the disconcerting preponderance of force arrayed against them, the French doggedly defended the houses against the attack of the sharpshooters of the Light Company and the Rangers. On their left flank their Indian Allies fought gallantly from the two groves of trees against an outflanking movement by twice their number of Rangers. Finally, stung by the accurate fire of the Rangers, the Indians rushed from the trees and engaged in a fierce hand-to-hand combat with the Rangers which ended with them being overwhelmed with 14 dead and a Chief and 4 braves captured. The Rangers lost 4 men killed. In spite of casualties, the Light Company advanced smartly against the Royal Roussillon Line Infantry Company in the houses on the right of St Nathan's. After a heavy fire-fight, the two forces met in hand-to-hand combat in the houses. It surged backwards and forwards until the French were all captured after losing 3 men who had been killed. The Light Company had 6 men killed during the advance and the mêlée.

The Grenadier Company of the Royal Roussillon Regiment defended the left flank houses against Roger's Rangers, whilst keeping alert for an attack by the British Line Company from St Paul's. On the 6th move, the British moved out threateningly but did not come within musket range. Encircled and having lost 7 men killed, the Grenadier Company surrendered on the 8th move, bringing the battle to a close. All in all, it was not a particularly bright performance by the French.

Whilst this small affair was going on, map-moves had continued and on the 4th move a contact was made in the Fort George area on the British left flank. Now the French plan of campaign became clear. Hoping to hold St Nathan's with a minimal force, they had thrown forward their full strength (with the exception of two companies garrisoning Fort Orleans) in a fierce attack on Fort George, garrisoned by only two British Line Companies. Obviously the French had thrown everything into a single attack and quite rightly the British Commanders agreed to amend their original plan of campaign. By relieving Fort George and destroying the French invaders the campaign would turn in the British favour as they could follow the French back to Franceville via Fort Louis and thus force the frontier at the same point as originally intended but from a different direction. The previous terrain of ruined villages and groves of trees was hastily cleared from the war-games table and the terrain for the second battle was erected over the area marked on the map.

At this point the critical might, with some justification, ask how the British commanders at Fort Robert, the captured villages in the centre of the map, and at Kingstown had become aware of the French attack on Fort George? It is a good question that can never be answered to the satisfaction of the more serious war-gamers in our ranks. This was a 'fun' campaign fought for enjoyment with the map and the movements upon it serving merely as an essential precursor to getting soldiers on a table-top terrain and fighting a battle or battles. It has been tacitly agreed by all concerned that whatever occurred at any point of the map would be known by some psychic or extra-sensory-perception gifts!

Having digested all that, the reader will find no surprise in the knowledge that from move 5 onwards the Grenadier Company and five companies of the 37th Foot were moving smartly up the road towards the crossroads from Fort Robert whilst the cavalry engaged in the St Paul's and St Nathan's operation, seeing that they were unlikely to be required, had begun to move down the road towards Kingstown. One company of the 37th were left behind to garrison Fort Robert.

The complete map of this campaign covered an area of 24 squares by 10 squares (each square being 1 in. by 1 in. and scaled so that 1 in. on the map equalled 1 ft. on the war-games table). The table on which the St Nathan's and St Paul's affair was fought formed an area on the map 8 squares by 5 squares with the crossroads at its exact centre. The terrain that was now laid for the attack on Fort George occupied the bottom left-hand third of the map, with its entire western quarter being formed of the waters of Lake Onogo stretching across as far as Kingstown in the bottom eastern corner of the table. Initially the map consisted only of roads, the lakes, the ruined villages, the forts and the two base-towns. It was our intention to insert trees and other topographical features in balanced positions on a table when contact was made in any specific area. This was done for the Fort George attack and, as the map shows, three groves of trees were placed along the eastern border of the map with a small hillock crowned by a clump of trees at the near-centre of the table.

Contact in the Fort George area being made during the 4th game-move, action commenced on the 5th move, when the British look-out reported the surface of the lake dotted with a fleet of 6 Indian canoes and 2 boats with 3 rafts in tow. With their attention focused on the invasion from the lake, the British garrison allowed a force of 1 squadron of Lifeguards and a troop of Burgundy Gendarmes to slip past on the road and make their way towards Kingstown. With the cavalry were a large party of Indians who settled into positions around the north and east side of the log fort to begin a ragged musketry fire. The French cavalry moved down to position themselves in the grove of trees just west of Kingstown with the idea of intercepting any attempts at relief by the 1st Regiment of Foot who garrisoned the Base. The canoes, boats and rafts ground ashore south of the fort and landed 3 companies of the Royal Roussillon Regiment with a field gun.

Fort George was unassailable from the river because of the rocks and bluffs on which it was built so the French infantry, out of musketry range but under fire from the fort gun, marched round to the eastern face of the fort in which its gate was situated. Meanwhile the Indians spread out along the southern wall and began a heavy and consistent musketry fire. From the north, the remaining two companies of the Royal Roussillon Regiment came marching down the road to position themselves in reserve behind their Indian allies who were firing at the northern wall.

The fort's garrison of 50 men and a gun was considerably outnumbered by their besiegers who consisted of 100 Indians and 125 infantrymen. Nevertheless they acquitted themselves well in the battle that raged for 11 game-moves, continuing without cessation throughout the night. On at least six occasions the Indians surged forward to clamber up the wall, sometimes they gained a footing but generally they were repulsed with considerable loss. The French infantry at the eastern wall made numerous attempts to batter the gate down with a large tree trunk which required half a company to carry, whilst their comrades kept up a heavy covering fire. The defenders' numbers gradually decreased and when on the 13th game-move, the Indians finally held the south ramparts, the garrison turned their gun to face inwards from the east wall and blasted them with grape shot at close range. On the 14th game-move, the sorely battered gate gave way and allowed the French infantry to enter, their first wave being wiped out by fire of the defenders who had massed their now scanty numbers to meet them. For the last two moves of the battle, the defenders could be counted in single figures but they had the satisfaction of knowing that they had reduced the original number of their attackers by two-thirds.

The fort fell on the 16th game-move with the garrison wiped

out to a man—they asked not for quarter and none was given. But their sterling defence, completely upsetting the French time-table, seemed to have won the day because the British cavalry from St Paul's had arrived at Kingstown at the end of the 10th game-move and from the 11th game-move onwards had been hotly engaged with equal numbers of French cavalry to the south of the grove of trees that lay near Kingstown. When the cavalry passed through Kingstown and out on the Fort George road, the 1st Regiment of Foot, the Kingstown garrison, marched out behind them but so confined was the area in which the cavalry mêlée was taking place that the infantry were forced to remain packed on the road as idle spectators for at least two game-moves. With hindsight, it might have been better for them to have marched out to the south and west so leaving the grove of trees on their left and attempting to outflank the cavalry and come to grips with the besiegers of Fort George. On the 14th game-move Roger's Rangers arrived at the top righthand corner of the war-games table and began to come onto the field south of the upper grove of trees. At the same time, the 37th Regiment's Light Company came onto the table at the centre point and moved westwards to the centre grove. Some two or three moves earlier, the French had detached the Indians who had been assaulting the north wall of the fort to counter this threat. These Indians had moved eastwards and were positioned in the groves of trees and around the small hillock in the centre of the battlefield. A running fight between these Indians, the equally nimble Rangers and the well-trained Light Company rippled back and forth along the eastern edge of the war-games table for a number of moves. As in all stages of this campaign, the Indians fought doggedly and had to be destroyed before any advance could be made.

At the end of the 15th game-move, the five Line Infantry Companies together with the company that had fought in the St Nathan's and St Paul's affair, arrived at Kingstown and joined the dense mass of men and horses seeking to push their way up the road from Kingstown to Fort George. The French cavalry, now reinforced by French infantry moving down from Fort George, were fighting doggedly but were being slowly pressed back.

Whilst this was an interesting exercise in war-gaming, it was now apparent that the French, much reduced in numbers, had no chance of attaining their objective of forcing the frontier and

K

capturing Kingstown. Not only were they greatly outnumbered by the oncoming British but, far from forcing the frontier, they were fighting desperately to hold their own and survive.

At this point, the British commander gently mentioned that the French and Indians who had been attacking Fort George, and were now hotly engaged on the Kingstown road, had been fighting without a break for about 15 hours and surely, with no chance of achieving their objective, they should be sensible and go home? The French commander hotly replied that whilst he admitted this to be the case, the British force with whom he was fighting had marched right across the map, taking the same period to do so! However he, the French commander, realised that his gamble had failed and he ordered his forces to break off the combat and retreat back to the lake. The British, equally weary, allowed them to go without pursuit and finally what was left of the Indians and infantry embarked in a far less number of canoes and boats than they had required to come southwards. As they went, the French commander darkly muttered something about '. . . there'll always be another day . . . it doesn't always get dark at six!'

*Technical notes*

On the war-games table, the Indian canoes moved 12 in. on the lake; the boats moved 6 in.—when towing one raft 4 in., when towing two rafts 3 in.

The single artillery piece possessed by the French and the British was a 9 pdr. Each of the forts was armed with two 12 pdrs. and each boat had a small 4 pdr. in its bows. During the description of the attack on Fort George, only one gun has been mentioned—this is because the guns were more or less permanently positioned so that one covered the lake whilst the other covered the gate. After initial firing on the canoes and boats, the gun pointing out towards the lake was unable to play any further part in defending the fort. The effectiveness of these 12 pdrs. was as follows:

A canoe sank at once if hit (throw for occupants 6's only save).

Two hits on a raft sank it; for a single hit throw for the numbers of men scored on a dice (6's save).

Three hits sink a boat otherwise throw for numbers of men scored on a dice (5 or 6 saves).

The 4 pdr. guns in the bows of the boat had no effect on the stout log walls of the fort, being primarily for defence against other craft. However, had it been possible they would have been useful in clearing away riflemen from the ramparts but once having beached, the French failed to position them so that they could be used for this purpose.

The French 9 pdr. gun, although firing regularly upon the fort, was singularly ineffective and when it was knocked out by a 12 pdr. ball halfway through the battle, the French appeared to feel no sense of loss.

The battering ram consisted of a stout piece of twig held in position between two files, each of five men, who moved forward at half speed. As it could only be successfully manipulated by these ten men and their losses were frequent, it took seven moves for the battering-ram to finally smash in the gate. The method of operation was for the log to be allowed 'two smashes' per move and a dice to be thrown for each 'smash'. The gate had a 'defence value' of 24—whilst the log was wielded by ten men every point scored on the dice counted, but a point was deducted from the dice for each man less than ten manning the log. Thus the total of 24 was only achieved after seven game-moves, for three of which the log was out of action because less than five men remained to wield it.

To successfully keep even a relatively simple campaign such as this in correct chronological order, it is essential to record everything on a War Diary. Although considered in some detail elsewhere in this book, the reader is reminded that this need not be an elaborate record but can consist of a sheet of lined paper with an inch margin on the left. In this margin, during this French and Indian campaign, the move numbers were noted but if the War Diary is covering larger movements of a campaign spread out over the map of a country for example then the number of the Day will be entered in this column. In this particular campaign, the Diary was headed 'The French and Indian War 1760'.

'1 day equals 12 game-moves' and then on the form proper, 'Move 1' in the left-hand column and then under 'EVENTS' 'Contact at Crossroads near ruined villages of St Nathan's and St Pauls'. Then followed details of the strengths of the respective forces and at day number 4 'Contact at Fort George'. Then moving down to move number 8 'French surrendered at St Nathan's' and so on. It should be remembered that this is the British War Diary that is being quoted—the French had their own War Diary and the two placed together at the conclusion of the campaign present an interesting picture of the lines of thought and subsequent events as seen from both sides of the war-games table.

This was a very colourful and interesting little campaign that ran over a period of eight hours in two evening sessions. Three war-gamers a side took part, but it could equally have been handled by one a side or as a Club Project with one war-gamer controlling a Company, band of Indians, squadron of cavalry, the Rangers, etc., etc. As with most of the other campaigns fought by the author of this book, the maps (two maps are always made—one for each side) have been filed away to remain as a ready-made area for another occasion when the French decide to chance their arm again, or the British make up their minds to force the frontier and capture Franceville once and for all.

# 18

# A Corps Campaign
## (Napoleonic)

Draw a map of any chosen area of ground, either real or imaginary. Prepare it to the scale of $\frac{1}{2}$ in. to 1 ft. on your war-games table and make the map large enough to be able to fit four of your war-games tables across it by five deep. Thus a map for a table 8 ft by 5 ft. will be 16 in. across by $12\frac{1}{2}$ in. Divide the map into one inch squares, each numbered in the top left-hand corner so that the top line of the map has squares numbered 1 to 16 then the second line starts again on the left-hand side 17 to 32, the third line 33 to 48, etc.

Next, cut a transparent template out of plastic large enough to cover, on the map, the area of your war-games table. For our map, it will need to be 4 in. by $2\frac{1}{2}$ in. Square it off in 1 in. squares.

*Narrative*

British forces are advancing into France in 1813 and in this sector their objective is to force the French troops out of any prepared positions they may occupy. Each side has the following force:

 3 Light Regiments (Rifle regiments and Tirailleurs).
18 Regiments of Line Infantry.
 6 Cavalry Regiments.
 5 Horse guns and
 5 Field guns.

Both the French and British Generals may divide their forces into three corps of roughly the same points strength (see elsewhere in the book for a table of points values for various troops). Next, nominate each corps an Assembly Area.

*British*

    Area number 1—squares number 17, 18, 19, and 20.
    Area number 2—squares number 21, 22, 23 and 24.
    Area number 3—squares number 25, 26, 27 and 28.
    Area number 4—squares number 29, 30, 31 and 32.

*French*

Area number 1—squares number 49, 50, 51 and 52.
Area number 2—squares number 53, 54, 55 and 56.
Area number 3—squares number 57, 58, 59 and 60.
Area number 4—squares number 61, 62, 63 and 63.

Both sides *must* assemble a corps in each of three of the four areas. Thus if the French assemble in areas 1, 2 and 3 and the British assemble in areas 2, 3 and 4, then each will have an unopposed flanking force which will be moving (on the map) whilst the two French and two British corps opposing each other are fighting. These map moves will be scaled to bear a relation to the table-top moves. They will be as follows:

Line Infantry and Field Artillery—$\frac{1}{2}$ in. on the map for every game-move on table.

Cavalry—1 in. move on the map for every game-move on the table.

Add 25% bonus for map-moves when moving on roads.

*Laying out*

For the first battle, it has to be decided if both sides are advancing to find each other or whether the French are setting up defensive positions. It must be borne in mind that the British, generally speaking, are advancing. If it is the former (both sides advancing towards each other) then it is an 'Encounter battle' and it will be necessary to decide if both armies lay down on the baseline of the table; 12 in. from the baseline; or 12 in. back from the curtain laid down in the centre of the table. If not advancing, then the French initially will be disposed in defensive positions on the table-top and the British will advance from the baseline.

*Setting-up*

As an example, let us assume that by a dice throw or some other means it has been decided that the first battle will take place in area number 4. It so happens that both the British and French have forces in area number 4; had one side not disposed a force in area number 4, then both sides would have been aware that they were overlapped—one on their right flank and one on their left. The French have decided to stay put in their defensive positions so that the British will move forward and their baseline will be the top of squares 45, 46, 47 and 48. If the top of the plastic template is now placed along the top line of those squares it will extend to downwards to cover

squares 45, 46, 47 and 48
61, 62, 63 and 64

Top half of squares 77, 78, 79 and 80.

This is the battle terrain and will be reproduced on the table-top.

### Further developments

If the British win this battle, they may select the next battle terrain in the following manner:

1. The defeated French may withdraw back straight or diagonally, any distance they choose but not more than $2\frac{1}{2}$ squares measuring from the baseline they occupied during the last battle (the top of the bottom half of squares 77, 78, 79 and 80).

2. The British transparent template is then placed on the map in any logical position adjacent to this battle area but not further forward than with its front edge on the top of squares 125, 126, 127 and 128 or a similar line. The covered area is then the area of the next battle.

3. The British who have just won may move forward to occupy the ground they have just captured and may remain there. The remainder of the British forces may then select their next battle area. This may be *any* $8\frac{1}{2}$ squares covered by the template *but* no further forward than line 113 to 128 and is dependent upon any French resistance in any of the areas over which this forward move might take place. This new area may overlap the area already used if desired—it does not strictly have to conform to the original assembly areas.

4. If the French win the first battle then the defeated British may retreat up to 2 backward or diagonal moves and the French may move forward to contact them again. Should they not wish to do this, then the British who have lost will retreat as above and the remainder of the British forces may select their next battle site as explained in para. 3. This will be the pattern after each battle. The specific situation will decide whether the next battle is to be an 'encounter' or a 'set-piece' battle.

### Attrition

If desired, losses in man-power may be carried on from one battle to another as a result of casualties. This may be decided in any way convenient or desirable to the opposing generals.

Should attrition not be desired then, after battle number one, both forces will re-group as outlined above in their original strengths.

When studying this campaign prior to writing it up for this book, the author recalled with nostalgia what actually transpired as it is written down on the movement chart for the campaign. Apparently, although area number 4 was the one selected by dice for the first battle, the advancing British did not have a force in that area. This meant that the French right flank force were unopposed and, at the same time, the British right flank force were in the same happy position. A further dice throw decided that the initial battle would be in an area where the French were disposed in squares 73, 74, 75 and 76, settled in defensive positions, whilst the British advanced from squares 41, 42, 43 and 44. At the same time as this battle began, the French right flank corps pulled out and began to move back across the map behind their other corps to cover their relatively open left flank. The British right flank force began to move forward to take advantage of the open area in front of them. Both forces moved on the map at the rates specified above.

Battle number one apparently took seven game-moves and at move number eight on the movement chart there are the following remarks:

Battle number one won by the French who draw back to squares 88, 89 and 91. The British retreat back to squares 24, 25, 26 and 27.

The French right flank force have reached an area covered by squares 103, 104, 105 and 106. The British right flank force have their Cavalry in square 132 and their Infantry in squares 99, 115 and 131.

Battle number two took place in squares 37, 38, 39 and 40, 53, 54, 55 and 56. Top half of squares 69, 70, 71 and 72.

It would seem that, by the start of battle number two, the French left flank force were already outflanked.

This battle, according to the movement chart, was won by the British in 5 game-moves. As a result of this defeat, the French re-grouped in squares 86, 87, 88 and 89 whilst the British moved forward to squares 70, 71, 72 and 73.

At this point on the movement chart a total of 14 game-moves had been used up.

Battle number three commenced on the 15th game-move,

when the French corps which had moved across from their right flank and positioned in squares 101, 102, 103 and 104 turned to face their rear where the British were assembled in squares 117, 118, 119 and 120.

This battle, which took 6 game-moves to decide, was won by the British. At this stage the French general decided that things were much too black and conceded the campaign. The author recalls this with some anguish as he was the French commander!

| *Move No.* | *Events* | | |
|---|---|---|---|
| 1 | French Left Flank Force in squares 73, 74, 75, 76 | | Battle No. 1 |
| | British Left Force in squares 41, 42, 43, 44 | | |
| 2 | Battle 1 | British Right Flank Force begins move | French Right Flank Force begins move |
| 3 | ,, | | |
| 4 | ,, | | |
| 5 | ,, | | |
| 6 | ,, | | |
| 7 | ,, | | |
| 8 | Battle 1 won by French Left Flank Force who draw back to squares 88, 89, 90, 91 | British Left Force to squares 24, 25, 26, 27 | |
| 9 | Battle 2 in squares 37, 38, 39, 40, 53, 54, 55, 56, 69, 70, 71, 72 | | |
| | British Centre Force in 37, 38, 39, 40 | | |
| | French Centre Force in 69, 70, 71, 72 | | |
| | French Right Flank Force reach squares 103, 104, 105 and 106 | | |
| | British Right Flank force in squares 99, 115, 131. Cavalry in 132 | | |

| | | |
|---|---|---|
| 10 | Battle 2 | |
| 11 | ,, | |
| 12 | ,, | |
| 13 | ,, | |
| 14 | ,, | won by British Centre Force, who move to 70, 71, 72, 73<br>French Centre Force re-group in 86, 87, 88, 89 |
| 15 | Battle 3 | French Right Flank Force in 101, 102, 103, 104<br>British in Right Force (outflanking movement) in 117, 118, 119, 120 |
| 16 | ,, | |
| 17 | ,, | |
| 18 | ,, | |
| 19 | ,, | |
| 20 | ,, | |

# 19

## Two Armies against One
### (Napoleonic)

It sometimes occurs that three war-gamers wish to take part in a campaign or else one has three new armies and is eager to use them all. The latter was the case when the author fought the following campaign, using a French Army and an Austrian Army against a British Army in about 1813. A map was made measuring 16 in. by 10 in., divided into sixteen squares each 4 in. by 2½ in., each of these squares was again in turn sub-divided into ½ in. oblongs. To the scale of ⅓ in. to the foot, this map represented sixteen war-games tables—each of the larger squares being a complete 8 ft. by 5 ft. war-games table.

*Map-moving*
Infantry and field guns—1 ft. per move (9 ft. per day).
Cavalry and Horse Artillery—1⅓ ft. per move (12 ft. per day).
1 day=9 1-hour moves. No night-moving.
On roads—add ⅓ distance.
On slopes and hills—move at ½ speed.
Rivers take 1 ft. to cross.

*Narrative*
A French force is coming from the South, by the most eastern road.

An Austrian force is moving from the South, up the most western road.

A British army is known to be moving from the North by an unknown route.

The Allies (France and Austria) are seeking to trap and destroy the British whilst maintaining their own line of communication. The British are living off the country.

*Method*
Initially, it is necessary to select the armies for the first moves or encounter. This may be done in the usual manner or in the fashion carried out by the author during the actual fighting of this campaign. On this occasion it was felt that uneven forces

(not necessarily to the choice of their commanders) would add zest to the campaign. With this in mind, points values were given the units that formed our armies.

A Line Infantry Battalion—4 points.
A Guard Infantry Battalion—5 points.
A Heavy Cavalry Squadron—3 points.
A Light Cavalry Squadron—2 points.
Rifle Regiment or Tirailleurs—2 points.
Guns—1 point each.

Each British regiment was then written down, by name, on a separate card; cards were also made out separately for each gun. The same was done for the French and the Austrian forces. Before the campaign commenced each general drew enough cards to total 20 points. After drawing, each general had a force, the substance of which was unknown to his enemy, which was not of his own choosing and might well contain a preponderance of cavalry and no artillery, for example.

Next, the map-moving began. At dawn on the 7th September, 1813, the British moved forward on their map. At this point it is worth mentioning that only one map was used so that all parties were aware of the opening movements of their opponent. After the British had moved forward, each of the Allies threw a dice and began to move that number of moves after the British in accordance with the number shown on the dice. For example, if the British moved 1 move in and then the Austrians threw a dice and scored 1 and the French scored a 6—then the British could move 1 more move before the Austrians came on to the map and the British could move 6 moves in all before the French markings were made on the map.

At nightfall, after 9 British moves (9 hours) all forces halted and consolidated their position. At dawn, all forces within 1 war-games table range of each other engaged.

The army that was obviously most suited to defence lay out in defensive positions up to the half-way line, whilst the other army rested on its baseline. There was no obligation for an army to set up defensively, if desired both armies could manœuvre from a baseline.

The battle was fought for 1 'day' (9 moves). The loser had to move completely out of the square on the map in which he had been defeated into a table immediately adjoining the one just used. Then map-moving continued until a further contact was

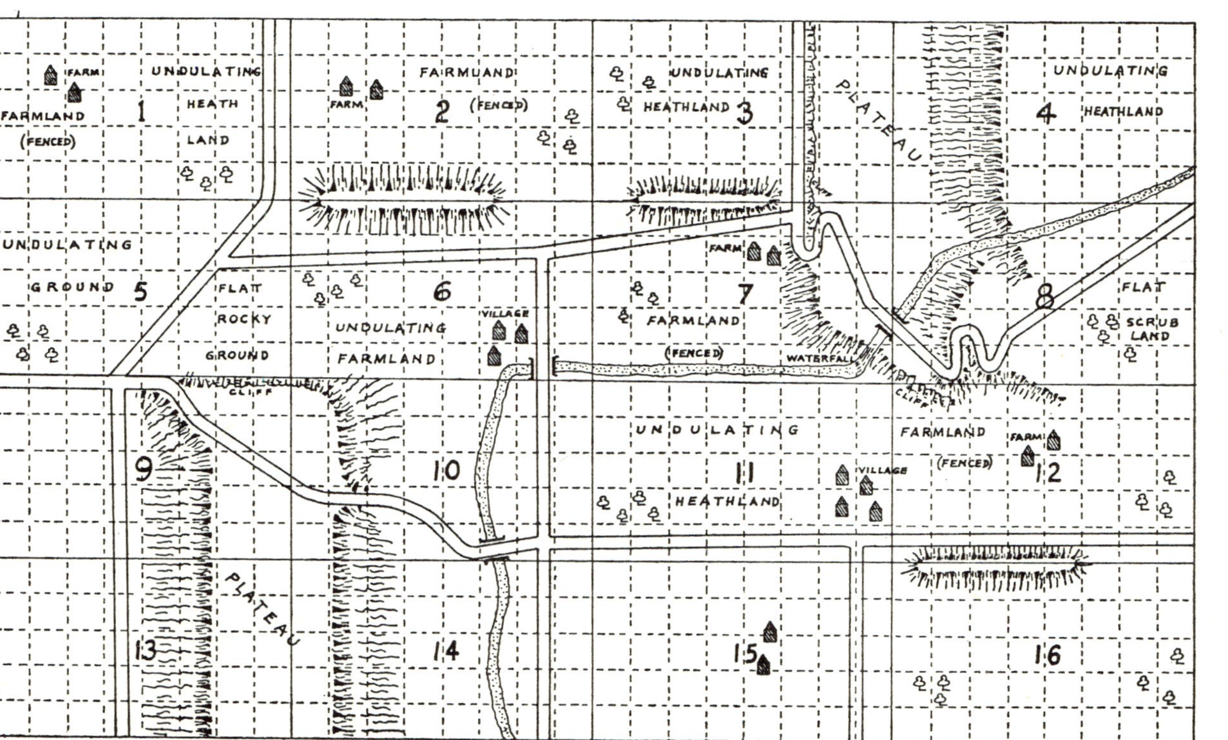

FARM
FARMLAND (FENCED)
1
UNDULATING HEATH LAND
FARM
FARMLAND
2 (FENCED)
UNDULATING HEATHLAND 3
PLATEAU
UNDULATING HEATHLAND 4
UNDULATING GROUND 5
FLAT ROCKY GROUND
UNDULATING FARMLAND 6
VILLAGE
FARM
FARMLAND (FENCED) 7
WATERFALL
CLIFF
FLAT SCRUB LAND 8
CLIFF
9
PLATEAU
10
UNDULATING HEATHLAND 11
VILLAGE
FARMLAND (FENCED)
FARM
12
13
PLATEAU
14
15
16

made, when both sides again drew for armies and carried on as before.

As there were three armies and only two could fight at a time, the third unengaged army were able to move at its specific rate on the map, and if time permitted could enter the battle. It had to give notice of its arrival when 1 map-move from the border of the war-games table.

Obviously, the British were endeavouring to defeat each of the Allied armies in detail, without allowing them to combine. To a certain extent they were able to do this and the campaign was carried on with roughly alternate battles between the British and the Austrians and the British and the French until at last weight of numbers prevailed and the British were penned up in the top left-hand corner of the map with the French to their front and the Austrians on their flank. At this point they conceded the campaign.

20

# A Narrative Battle

(France v. Austria and Bavaria in Italy 1859)

Quite frankly, this campaign was one of complete self-indulgence.

Every single one of its many features was so designed as to allow the author to indulge himself in long-cherished dreams and schemes, to blatantly follow well-trodden paths through favoured military scenes and periods and to generally round off and tidy up his general war-games scene by utilising figures and other items that had been lying around for a long time in aggravating disuse.

The Horse-and-Musket period has always attracted the author more than any other so that his battles and campaigns have ranged from the English Civil War through Marlborough's campaigns, the Seven Years War, the Napoleonic Wars, the American Civil War, the Franco-Prussian War and British Colonial campaigns. At the same time there was always a consciousness that there were many smaller wars that held great attraction for reasons of the armies involved, the terrain over which they were fought and the weapons and equipment that was used. Some of these conflicts were obscure to the point that it was difficult to even obtain much information about them—to contemplate converting or casting and painting table-top armies of the forces engaged in the Carlist Wars or the Italian Wars of Independence was too time-consuming to be even contemplated. Nor was there any attraction in fighting these wars with figures from other campaigns or periods. It was a problem with which the author wrestled on many occasions when he should have been sleeping, reading, writing, casting, converting or painting!

Like many other innovations and projects contemplated by the enthusiastic war-gamer, the solution arrived by a round-about route. In the course of conducting mid-19th century Horse-and-Musket campaigns, three principal armies had been accumulated—British, with a Home full-dress establishment and Colonial with the appropriate dress and equipment; French, both for the Franco-Prussian War and for the North African Campaigns; Prussian, which included smaller groups for Bavaria, Saxony, Hesse, and Württemberg. The Franco-Prussian War has been re-fought on numerous occasions, the accuracy and realism of its war-games rules and background being stressed by the fact that the Prussians invariably triumphed. The Colonial forces of the British, French and Prussians clashed in a mythical Colonial Campaign in an equally mythical continent of East Asia. But always there remained a nagging consciousness that there were no real-life wars between the British and the French or Prussians during the mid-19th century whilst the presence of large armies of three major Powers represented an unbalanced situation. It seemed obvious that a fourth fairly large national army was required, either to ally itself to one of the other three or to go ahead on its own account and open hostilities with one of the already existing armies. The Russian Army and the Crimean War came most readily to mind but quite frankly their performance seemed so pathetic; with the possible exception of the Cossacks, their regiments appeared uninspiring so that the author could not contemplate leading them with any confidence nor fighting against them with much interest.

With the problem still in mind, the family proceeded to Innsbruck in Austria where a 'must' was a visit to the Kaiser Jager (Imperial Rifles) Museum. It is an inspiring place chock-full of uniforms, arms, prints, plates and paintings connected with the Austrian Army in general and this famous Tyrolean Rifle Regiment in particular. Within minutes the problem was solved—it was to be a mid-19th century Austrian Army that would form the Fourth Force! Hours were spent in the Museum drawing and taking notes so that on returning to England everything could be pushed to one side whilst the Austrian Army came into existence.

Now came the first real pandering to the author's whims. Having never felt really happy about the sight of model soldiers advancing across the battlefield whilst kneeling and firing or standing and firing, or remaining in reserve in a flagrant running position, the author has always believed that the ideal position for war-games figures would be one of the following:

(a) Advancing, with the rifle and bayonet held out in front.
(b) Advancing, with the rifle and bayonet held across the front of the body at High-port.

# OPERATION SHEET

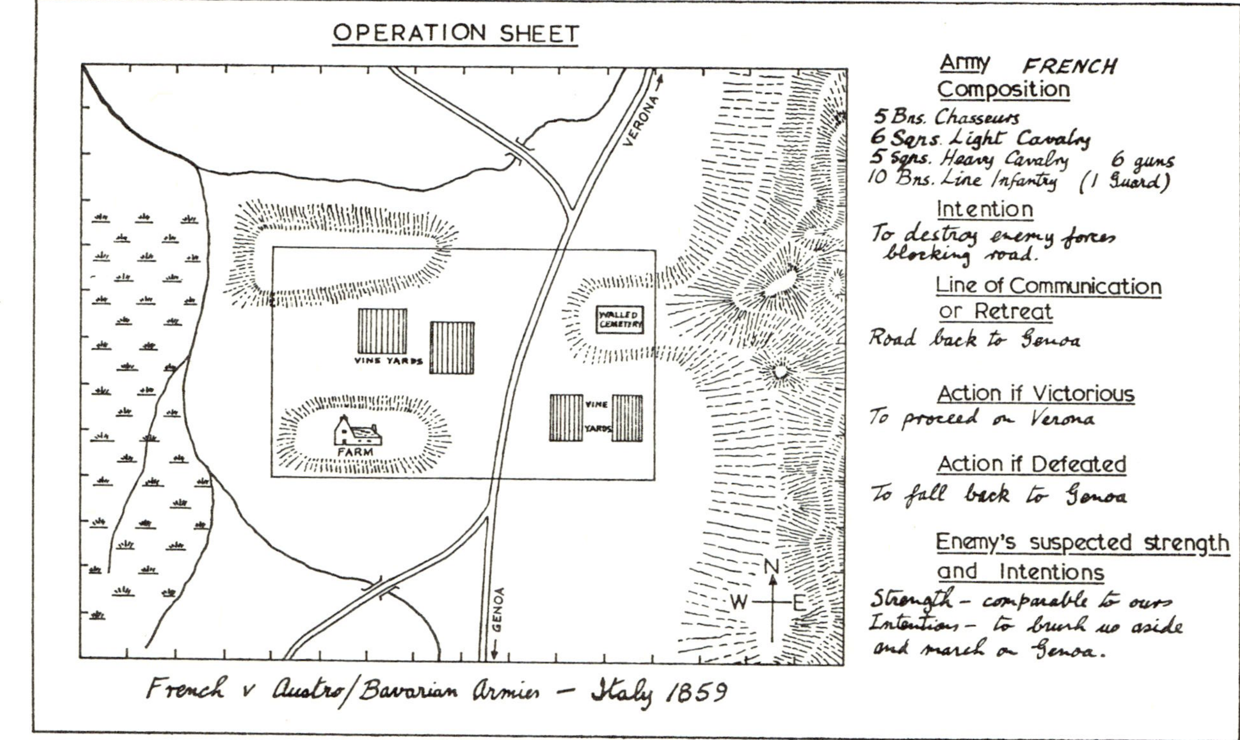


French v Austro/Bavarian Armies — Italy 1859

(c) Standing with the rifle and bayonet held out in front in the 'on guard' position.

To these positions is added a proviso that a marching figure is considered admissible but only for the more formal and stately periods of warfare such as the Marlburian Campaigns.*

Unwilling to give an inch to prevailing circumstances, the author set about building up an Austrian Army in 30 mm scale when so far as he was aware there were only three known figures of this type and scale in existence. One of them was a Swedish African Engineers (S.A.E.) figure that had been out of production for some ten to twelve years whilst the other two were American made and at that time virtually unobtainable in this country in any quantity. Converting, moulding and casting occupied many months to finally result in 200 Austrian Line Infantrymen, 100 Austrian Kaiser Jager plus 4 guns with crews and a General Staff. Digressing, the resplendent uniforms of these generals were taken from the sleeve of an L.P. 'Militär-märsche aus Osterreich' purchased at Innsbruck. But to have 350 gleaming, cleaned-up figures is not to have an army, as war-gamers well know. Providentially, the Olympic games arrived and the author managed to paint almost the entire army whilst staying up all night in front of a television set watching the athletics. In fairness it must be confessed that the Austrian Infantrymen wore a practically all white uniform whilst the Jager are all grey so that the task was simplified by spraying en masse the infantrymen with white and the Jagers with grey, so that only facings, etc., had to be painted on.

So there we were with our Fourth Force—where and whom were they to fight? Again, the answer arrived in a roundabout manner when the author was nostalgically studying some post-cards purchased in Innsbruck depicting incidents in the Battles of Magenta and Solferino, both fought in 1859 during the Italian War of Independence against the Austrians. For a brief horrifying moment it appeared as though it would be necessary to make and paint an Italian Army but commonsense prevailed with the realisation that the principal opponents in this war were the French and the Austrians—both of whom were in plentiful supply on the author's shelves.

For war-gaming purposes the choice could hardly have been

* All the figures for this Austrian Army were to be in the same position —that of (a) above. Different regiments were denoted, as in real life, by the colours of their facings and their banners.

more exciting. Here was a true mid-19th century Horse-and-Musket campaign fought between two countries noted for their gay uniforms and dashing tactics over a country personally known to the author through an arduous World War II trek from Salerno to Venice, taking in such interesting sight-seeing as Cassino, the Gothic Line and the crossing of the River Po. The armies were most stimulating as it was a period when Hussars looked like the chorus from a stage musical, when white-clad Austrian infantry fought red-trousered Frenchmen aided by Zouaves and Turcos in their reds, pale blues and yellows. The Foreign Legion could be there with the French Imperial Guard in their high bearskins fighting against green-uniformed Austrian chevau-leger with high black polished helmets. There was a great temptation to hold up the proceedings whilst making Italian troops from the states of Naples, Sardinia and Sicily—the Bersaglieri with their cocked plumed hats and the red-shirted, white-havelocked soldiers of Garibaldi were mouth-watering.

Only the slightest of delays occurred with the realisation that this small Austrian army were wildly outnumbered by the vast French forces accumulated during the years of their table-top wars with Prussia. What more natural than that the troops of that South German State of Bavaria should ally themselves with the Austrians in order to acquire a slice of Northern Italy? So the light blue uniformed, black-crested, helmeted soldiers of Bavaria with their Cuirassiers and lancers joined the Austrians in what was to be a most pleasant alliance.

For those to whom this war is but a series of Italian-sounding names, some details will be of interest. Divided into nine separate states the peninsula of Italy had been fighting since the 1820's to secure freedom from domination from Austria and to become unified into a single nation—we are concerned with the third attempt launched in 1859 with the support of France. King Victor Emmanuel II of Sardinia (Piedmont) formally enlisted the help of Louis Napoleon III of France and in Spring 1859 an army of Piedmontese and French pushed eastward across the Ticino River into Austrian-held Lombardy. Led by the French Marshal MacMahon, the Allies won a battle at Magenta in June—a bloody affair which caused young French artists idealistically serving in this war of liberation to discover the new colour of magenta to describe the appearance of blood staining the red baggy breeches of the French infantrymen.

Later in the month of June, the French and Austrians again clashed at Solferino, where Louis Napoleon III and Victor Emmanuel of Piedmont personally led the Allied armies. It was a prolonged and bloody battle with the French again victorious. The heavy casualties here and at Magenta three weeks earlier so sickened the French King that he signed a separate truce with Austria and left the war. A Swiss traveller, Henri Dunant, was equally sickened by the sights he witnessed and as a result helped to establish the International Red Cross in 1864.

So much for history—from a war-gaming point of view a revealing sight into the style and method of the campaign can be gained from Philip Guedalla's book *The Second Empire*. The 80-year-old Baron Henri Jomini, who had served with Ney at Ulm and Jena, drew up a plan for Louis Napoleon's use in Italy.

'It ignored completely the unauthorised innovation of railways, and it depended on its success upon the obliging courtesy of an enemy who would keep reasonably still. But since it was used against the Austrians, it was entirely successful, and the French enjoyed in 1859 the pleasing experience of defeating, with the methods of 1809, an adversary whose military thought was that of 1759 . . . but if the Austrians had been Prussians . . . the French would have been swept against the Alps.'

On both sides the services of scouts were so ill-organised that the Commanders of both armies rarely had any clear idea of the other's strength and whereabouts. From these sentences the war-gamer will gather that an Encounter Battle is most likely, with both sides fighting in their own inimitable way—the French with dash and élan and the Austrians with courage and doggedness but in a long-outmoded manner. Do not let the war-gamer who commands the Austrians be influenced by the reputed remark of a famous pressman of the time that 'The Austrian army exists only to provide victories for other countries'.

The story of self-indulgence continues with the confession that although admitting to the fascination, realism and interest of Map Campaigns, the author would rather get on with the real meat of the hobby—the table-top battles. However, generally speaking it is necessary to use maps to run a war-games campaign, the essence of which lays in the degree of continuity and connection it gives to battles and to fight a single battle with nothing before or nothing after considerably detracts from its interest. Additionally, being in the fortunate position of having a permanent war-games room in which table-top terrains can be left in position for weeks at a time, the author has at his disposal other methods of operation. Therefore it was decided that this Franco-Austrian campaign should be fought as one large 'Narrative Battle'. This consists of laying out a realistic terrain, probably taking more time and care than usual in its preparation, and fighting over it with the entire forces of both armies in a battle that might well last for four or five evenings, the essential thing being that a comprehensive story should lead up to the battle which should culminate with the attaining by one side or the other of a recognised objective.

The author is well known among his friends as a war-gamer who finds equal interest in building up realistic terrains and battlefields as in making the soldiers that fight over them. A chalked road or river pains him, whilst he finds it as painful to ascend a hill of regular polystyrene steps as he would the real thing. In 1859 this war was fought in Lombardy, a beautiful part of Italy embellished by a large number of widely differing topographical features all encompassed within a relatively small area—in other words, much like a real-life version of a war-games table. In accounts of the Battles of Magenta and Solferino one finds frequent mention of canals, rice fields, farms and villages on hills, fields separated by dense thorn hedges or rows of mulberry trees bound with wire on which vines are trained, houses built on terraced slopes, large lines of poplars marking roads, walled cemeteries and white churches. Indeed it was a terrain to gladden the war-gamer's heart but not one to be attempted lightly or hastily thrown together a half hour before the combatants arrive.

One of the most marked failings of the average war-games terrain lies in the absence of dead-ground, those hollows in which men are sheltered from fire or which prevent guns and small-arms fire from being directly trained upon them. Short of a sand table, the most realistic method of making a war-games table is to fashion green material over shaped projections so as to give the rise and fall of ground. But this is not as easy as it sounds because the material will crease and the slopes that descend from carefully positioned blocks of wood will sag or else will be too steep to allow our miniature warriors to stand upon them. After much experimentation, the author has settled upon a method of positioning shapes on the table-top, covering them with a layer of under-carpet felt upon which rests a thickish green-colured plastic furnishing fabric that can be purchased at most do-it-yourself stores. This material has the merit of allowing roads and rivers to be realistically painted upon it with poster paint, to be easily rubbed out with a damp sponge or cloth when a new terrain is required. Hedges and trees may be mounted upon green or brown Fablon, adhesive side downwards and firmly stuck into position on the plastic, to be peeled off when required. Italy is noted for its hilltop mediaeval white villages reached by winding roads—a most attractive and interesting piece of war-games terrain to be made by the enthusiast.

For this battle such temptation had to be resisted because with such large numbers of troops deploying on a table only 8 ft. by 5 ft. it was essential that the terrain was not too encumbered by hedges, houses and other features that might so hinder the movement of troops as to bog-down the battle. Eventually was constructed what appeared to be a simple terrain that was realistic in appearance and sufficiently open to allow for a reasonable degree of tactical manœuvring. An example of the tidying up that this campaign allowed was the author's delight in being able to utilise a cardboard cut-out model of a mountain chalet (purchased in Germany) backed with stiff cardboard and positioned on a small baseboard; this delightful-looking house formed a keypoint in the battle that was to follow. Another feature much fought over was an Italian walled cemetery characteristically denoted by four tall poplar trees stretching solemnly to the sky at each of its four corners.

### Narrative

In June 1859 the French/Italian armies were moving northwards from Genoa whilst the Austrian/Bavarian forces were marching southwards towards them from Verona. Because of their completely inadequate scouting systems methods neither commander knew of the whereabouts of the enemy. Furthermore, headlong collision was imminent because of the nature of the ground over which both armies were passing—to the east were the foothills of an extensive mountain range, to the west the wide and swift-flowing river bordered by extensive marshlands. The arena formed by these natural barriers was undulating ground, with some vineyards and a walled farm and cemetery crowning the tops of two ridges. Once having entered this natural arena, one side had to give way if the other was to advance—the only roads capable of bearing armies of this size were those that went northwards to Verona and southwards to Genoa. On this particular day, both commanders with unaccustomed military thought had thrown forward strong forces of light cavalry and riflemen together with some Horse Artillery. These forces will come on to the table and fight for 3 game-moves before the main bodies arrive. On the conclusion of the battle, the line of communication of both armies is the road down which they were originally marching. That is to say, the Austrians will retire back along the Verona road whilst the French will retire down the road to Genoa—in both cases there are side roads of inferior quality, both of which pass over bridges which would constitute a serious bottleneck in the event of a hasty retreat.

A 'Battle Sheet' was made out for both armies, containing details as shown together with a map of the battlefield and its surrounding area. The Battle Sheet details such vital points as the intention of each commander, lines of communication, the action to be taken in the event of victory or defeat. The use of this method ensures that a certain degree of reality is brought into the game by moving away from the common war-games failing of entering a war game without really having any set intentions and then closing it down at a point where one commander decides to concede whereas, in real life, his troubles would only be beginning as he sought to withdraw the remnants of his army behind a screen of cavalry and guns.

It has already been mentioned on numerous occasions that this campaign was an excuse for the author to test or try out numerous pet theories and ideas that had been fomenting within his war-gamer's brain for many years. Some of them worked—some of them failed but all agreed that they were interesting and in most cases worthy of further investigation, alteration or amendment. The first of such schemes necessitated purchasing some thick transparent plastic sheet (sold by Marley shops) cut to the size of 8 ft. by 6 ft. A length of cord was stretched lengthways along the table, about 3 ft. above the surface, over it the plastic sheet was draped so that it fell in double thickness down to the table-top. When standing on one side of the table with the sheet approximately 2 ft. 6 in. from one's face, it was just possible to see movements made behind it but extremely difficult if not impossible to detect what those movements actually were. In this manner an attempt was made to represent the degree of vision that troops would have when coming on to a battlefield with perhaps the complications of mist or the dim light of dawn or dusk. Both commanders laid out their light troops at any point they wished on their side of the curtain. Then, when both agreed they were ready, the plastic curtain was removed and the battle commenced. Both commanders had a slight idea where their enemy might be but neither knew the strength or the type of troops facing them. When one considers the various uses of plastic mentioned in this book, i.e. model soldiers, coverings for maps, the surface of war-games tables and now vision screens, one is tempted to pause a moment and offer up a hasty prayer of thanks for modern science!

Still using the campaign as a 'guinea-pig', advantage was taken of the fact that it spread over four evenings. Each of these evenings was used to test or demonstrate some particular theory or mode of war-gaming that had been milling around in the author's mind as a mental exercise but lacked the necessary practical test. The majority of the first evening was taken up by the Light Infantry and Cavalry probing and jockeying for positions that would give their respective main bodies the greatest possible advantages when they arrived. These light troops had more or less moved into the centre of the table by the time both baselines were filled with the massed ranks of Line Infantry, Heavy Cavalry and Field Artillery. Beforehand, two large maps of the area had been drawn to scale of 1 in. on the map equal to 12 in. on the war-games table (inch-squared graph paper was used). Each of these maps was divided into three equal longitudinal sections, each 20 in. apart. When the main bodies arrived on the baselines, they did not immediately pour down on to the table but were moved on the larger map and the troops only placed in position when they came within firing or charging range. This was an attempt to simulate the fog of war which it was considered would prevent a force on one side of the battlefield from clearly seeing its enemy on the other side —it was a continuation of the theory started by the use of the double thickness transparent plastic dividing curtain. All actual movement *on the table* was restricted to the centre 'action area'. Opinion was divided on the success of this method but a fair summing up would be that it is a little 'messy' because of the dual role of working on a map and on a war-games table, it is frustrating to see nothing in the two outer thirds of the table because in reality *some* movement, even the glitter of bayonets or the dust from the feet of horses or the wheels of guns, would

A closer view of the landing of the 1st British Combat Group.
(Chapter 30 'A Landing in Force'). Loaned by Brian Baxter.
(Crown copyright)

Close action in a village during the battle between 'Guard and Panzer Grenadiers'. (Chapter 31). Loaned by Brian Baxter. (Crown copyright)

*Opposite/Above*   The landing of the 1st British Combat Group. (Chapter 30 'A Landing in Force'). Loaned by Brian Baxter. (Crown copyright)
*Below*   The road leading to the bridge, showing a pillbox that held up the British forces. (Chapter 30). Loaned by Brian Baxter. (Crown copyright)

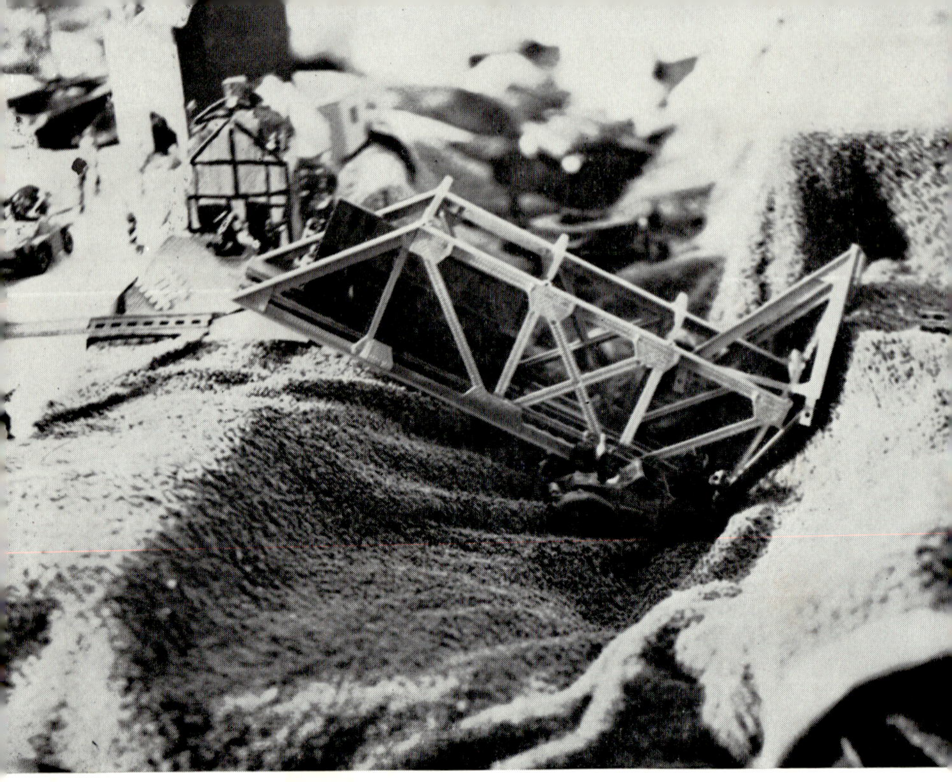

The battle around the railway bridge. (Chapter 31 'Guard and Panzer Grenadiers'). Loaned by Brian Baxter. (Crown copyright)

reveal the presence of troops. But more than that, it detracted from what must be considered the highlight of war-gaming—the sight of massed armies of troops bravely moving across a war-games table.

Yet another method was attempted on the following evening. The book *Advanced War Games* contains a detailed description of a method of war-gaming devised by American war-gamer Jack Scruby under the heading of 'Continuous Combat'. Although not to everyone's taste, and probably completely unknown to present-day war-gamers, there was the germ of reality about this method which would appear to justify further investigation. With that in mind, a provisional approach was made during the battle under discussion—it was not particularly notable but, like so many other innovations in this hobby, it might be with patience and the testing of time. Many good and worthwhile war-gaming ideas are discarded after one perfunctory trial when, with patience, tolerance and a little understanding on the part of one or both of the players, a very worthwhile contribution towards the progress of the hobby might have been achieved.

In this system, as might logically be expected, the Attacker moved first. He detailed on an order sheet the reasonable groups and supports that formed his force (i.e. right flank, centre and left flank, etc., etc.). Each of these groups moved as one and whilst one group was moving and in action, the other groups remained frozen.

At *any* stage of the Attacker's move, the Defender can announce that he is counter-attacking with his own units within reach of the Attacker's moving group. The Defender cannot pick one unit out of a group whilst it is moving but he must consider the group (including its supports) as a whole.

When an Attacking force is counter-attacked, other units of that army that have been frozen up to now may move in support.

Both Attacker and Defender must consider all situations according to the proximity in their respective move-distances of their other units.

Light troops (Jagers, Chasseurs, Tirailleurs and Riflemen, etc.) who wish to split-move (that is, move-fire-move) will do so at the start of the move.

When the Defender counter-attacks, having challenged the Attacker's moving group, then the system is carried on as already described but with the original Attackers now classed as Defenders. If there is any doubt as to who attacks or who counter-attacks but cannot be reasonably settled by amicable agreement then as a last resort a dice can be thrown for order of movement in that sector.

*Firing of artillery and small arms*—firing takes place *at will* by both Attacker and Defender. The firer announces his intention of firing and then carries it out. It may be before, during or at the end of troop movement.

If the target-unit also wants to fire then both sides dice for priority of firing.

Only one round of firing per unit or gun will take place during each game-move.

*Firing on Attackers coming in from a flank*—if a unit has flank guards out, then they take *half* the Attacker's move-distance to turn and face the direction from which the threat is coming.

If they have *no* flank guards out and are

(a) pinned frontally at the same time then this is an un-opposed (no firing) flank attack.
(b) if they are not pinned frontally, then dice throw and require 4, 5 or 6 to turn and fire as in the opening sentence of this section.

It takes *half* an Attacker's move for a gun to wheel 90°. It takes one complete Attacker's move for a gun to completely change position to a flank.

Just as it seemed that the third evening's war-gaming would be carried out using our known and tried methods, a suggestion was received from war-gamer Ron Sargent (late R.S.M. Hampshire Regiment). This suggestion seemed so promising that it was immediately acted upon and would seem to hold perhaps the best chances of success and adoption of any tried out in this particular battle.

The crux of Mr Sargent's argument lay in his belief that war-games figures, erect or splendidly horsed, lacked a tactical appearance on the war-games table. When the rules permit, one is frustrated on the occasion when fire is brought to bear on a line of infantry standing in a ploughed field only for the opposing general to triumphantly scream 'Ah! But they are lying down in the furrows!'

Of course, one could duplicate figures, having men in various positions for various situations (with the inexpensive availability of Airfix figures this could be more than a dream) but, as Mr Sargent says '. . . this can be costly even if models in the poses you require exist, to say nothing of the time consumed in painting them'. Instead, he suggests that a better tactical picture can be painted by using vari-coloured plastic tiddly-winks counters and giving each counter a tactical code. For example:

RED means the unit is firing (used for all arms).
GREY means the unit is lying down.
BLUE means the unit is running.
BUFF means the unit is crawling.
LIGHT BLUE means the unit is kneeling.
YELLOW (for cavalry) means they are dismounted, mounting or fighting on foot. For artillery it means unlimbering or limbering-up.
WHITE means the unit is loading (really for artillery but could be used for all arms).

When the forces are deployed, a counter is placed behind each unit immediately enabling both generals to see the tactical pose of the figure or unit concerned. These coloured counters will work for any period and, as Mr Sargent says, '. . . next time you engage those infantry in a ploughed field, unless your opponent has put a grey and possibly a buff counter behind them—he's had it!'

All in all, this turned out to be a most enjoyable and memorable campaign. In part this was due to the fact that a number of visiting 'firemen' took part in the various evenings' campaigning —including one from America, one from South Africa and two from varying parts of Great Britain (one of them was a serving Captain in the British Army but even his flanking attack with a cavalry brigade did not come off!). One of the most significant aspects of a war game lies in the fact that in later days one is unable to remember whether one suffered defeat or gained victory. This implies that the game was so enjoyable that the result was of little account. In this particular battle, maybe the French triumphed, or was it the Austrians? However, it was an interesting, unusual large-scale battle fought amicably and with relish by all concerned—and what more can one ask from a war game?

# 21

# Stonewall Jackson in the Shenandoah Valley
## (American Civil War)

Stonewall Jackson is in the Valley with a small force, to try and prevent the Federals sending any reinforcements to the Union Armies in front of Richmond. Lincoln has ordered that until Jackson's forces are destroyed, no Federal force will leave the Valley. The Union forces are exactly three times as strong as Jackson's force.

Under the overall command of Sheridan, the Federal force are disposed as follows:

Army of the Shenandoah (under Sheridan himself) at Winchester.

Army of West Virginia (under Crook) at Martinsburg.

6th Army Corps (under Ricketts) at Harpers Ferry.

*Each* of the forces totals 5 Infantry Brigades (10 Regiments), 3 cavalry squadrons and 4 guns.

Jackson's force—composed of 5 Infantry Brigades (10 Infantry Regiments), 3 cavalry squadrons and 4 guns—is lying at Staunton.

Winchester and Staunton are respectively the Federal and Confederate bases, through which their forces are reinforced and supplied.

It is necessary for the war-gamer to obtain or draw a map of the Shenandoah Valley, not a difficult proposition if one has almost any of the books of the Civil War. It should be drawn to a size of about 24 in. by 18 in. and scaled in one-inch squares, if possible each inch square should represent a 12 in. square on your actual war-games table. Thus, if your table is 8 ft. by 6 ft., your whole map will cover 4 tables by 3 tables.

As in real life, the strength of the Confederate force lay in its speed of movement, and that direction of movement coming as a surprise to the enemy, so that Jackson could strike one army, withdraw, march and then strike another one. In this campaign, the Confederate forces are considered to be 'Jackson's Foot Cavalry' and have an extra move distance over the Federal troops (see later).

To cover the other element of surprise, both generals have a map and move by mutual agreement until a contact is made or until a force is detected by cavalry scouts, etc. *But* the Confederate forces will move as *two* armies, one of which is the *real* force and the other a dummy—which is which not being known to the Federal general. When a battle-contact is made on the map, if it is the fake army then Jackson reveals this and the Federal force will not have a battle to fight. If the contact is between the two actual forces of Jackson's and the Union then a battle situation has occurred and a war game will take place.

A plastic, transparent template scaled to 1 in.=12 in. on the actual table is then placed on the map, a pin being pierced through its exact centre and stuck into the point of contact on the map—the area covered is the battle ground and is reproduced on the war-games table.

Jackson's sole intention is to keep his force intact and pin down as many Federal units as possible. When a battle is joined the distances that the other Federal forces are from the battlefield is worked out and, moving at the prescribed rates, they may move towards the battlefield and arrive at some stage of the battle, perhaps on a flank or even in Jackson's rear. Jackson may even break off the battle as soon as he finds he is outnumbered or even suspects that this might occur.

25% of the actual war-games losses are replaced after the battle through the two base towns, these reinforcements move at prescribed rates and may be affected if the line of communication between their force and the main army is cut by the enemy.

The Federals may attempt to reinforce troops. When battle is joined a courier is sent to their nearest friendly force—he may travel at the rate of twice the normal infantry rate. When he arrives at the force, a Chance Card is drawn to see what the Commander of that force will do. The force moves in accordance with the instructions on that card. (See details in *Advanced War Games*.)

Before the battle, wind direction is decided and if the friendly force are within 12 in. on the map in direction that wind is blowing, then they are considered to have heard noise of battle and to be 'marching to the sound of the guns'. But they still have to draw a Chance Card and march in accordance with its instructions.

*Map-moving*
Federal forces: Infantry 4 squares per day.
    Cavalry 6 squares per day.
    Guns 4 squares per day.
Confederate forces: Infantry 6 squares per day.
    Cavalry 6 squares per day.
    Guns 6 squares per day.

These moves are moves on the road, cross-country moves halved.

The advantage of this type of campaign lies in the fact that one only requires two relatively small-sized forces, because the Rebel's force is small, whilst the Federals, although having three armies on paper, actually only have the same number of men as the Confederates. A few extra men are required in case Jackson daringly decides to fight a reinforced Federal Force, but no Federal force is to remain larger than those sizes initially named any longer than for the one battle in which they might have been reinforced.

The 'fake' Confederate Army, having been revealed when a contact is made, is then at liberty to be moved not more than one day's march from the point at which it was unmasked. It can then begin moving again, together with its genuine companion army, as before.

Both generals should allocate a period for map-moving, perhaps an entire evening in the first place, to get the campaign under way. If this is not possible, the whole lot can be done by one man doing all the map-moving and only declaring positions and contacts to his enemy general on the actual night of the war game. This is done by having the fake army declared on a piece of paper and sealed in an envelope, the genuine army likewise being marked and sealed—both by an outsider, a third party. Thus, the general making the map-moves is moving as much in the dark as the Federals. When the fake force is unmasked, then the third party again obliges and seals the two forces in envelopes and off they go again!

The campaign ends when Jackson is thoroughly defeated, or when his reinforcements cannot get through to him, or when the three Union armies are so moved as to trap him, or when he is forced out of the Valley.

An interesting addition to the campaign lies in keeping a daily War Diary—assuming that the campaign begins on, say, 1st

April, 1863, and marking each day's move on the diary, contacts made on, say, the 7th April, and so on. Then, daily weather conditions can be inserted with effect upon movement on the map, and a prearranged delay period can be decided so that, by holding up the Federal forces in the Valley until, say, the 29th April, Jackson has won the campaign.

## What Happened to Jackson in the Shenandoah Valley

This campaign was fought with me 'guiding' the movements on my master-map, and keeping my opponent informed of events and letting him know when a contact had been made and a battle was due. Before I did the actual map-moving, my opponent indicated his general strategy and I followed that as much as possible. This method was used to avoid the 'wasted' evening when we both sat poring over maps, one of us on each side of the room instead of being upstairs fighting an actual battle. It would help the reader if he has in front of him a map of the Blue Ridge Mountains and the Shenandoah Valley area—available in almost every Civil War book.

The Federal Commander has got three armies, the Army of the Shenandoah under General Sheridan, the Army of West Virginia under General Crook, and VI Corps under General Getty. Each of these armies equalled Jackson's force in size, being 10 Infantry Regiments, 3 Cavalry Squadrons and 4 guns. First Federal orders were:

Army of Shenandoah to move *west* from Winchester, through gap in range, up west side of range, turn *east* through Southern Gap to seize Harrisonburg. Dependent upon the situation at that stage, it was then to either move north on the Strasburg Road, exploit towards Cross Keys and the Shenandoah crossings *or* split in order to perform both missions. During march to Harrisonburg, it was to disregard any clash between Jackson and other Union forces—Harrisonburg was its main objective.

The Army of West Virginia was to move *south* through Winchester on Strasburg, marching slowly so as not to outpace the advance of VI Corps. This Corps was to move from Harpers Ferry *east* of the Blue Ridge Mountains, heading south for Ashby's Gap. Half the Corps was to take up a defensive position in the Gap, the remainder to press on and take up defensive position in Manassas Gap. Once VI Corps is in position, Crook was to speed up his advance for Winchester, taking the road for Strasburg but being prepared to diverge towards Ashby's Gap in the case of Getty being engaged.

The general intention of the Union forces was to keep two armies holding the Northern and Eastern exits from the Valley and pressing Jackson frontally, whilst the third Army swung in on his rear to put him in the bag.

Jackson's Confederates (formed as one Real and one Fake army) leave Staunton—Army A towards Manassas Gap, Army B up Strasburg Road and through the town.

The Federals moved as follows—Army of Shenandoah left Winchester, moved south-west to arrive west of mountains in line with Strasburg. Army of West Virginia leave Martinsburg, move slowly towards Strasburg, where they make contact with Confederate Army B. VI Corps leaves Harpers Ferry, arrives at Ashby's Gap, leaves half force there (6 Infantry Regiments, 2 Cavalry Squadrons and 2 guns) whilst remainder under General Getty move into Manassas Gap, where they are subsequently attacked by Confederate Army A.

(Sealed envelopes opened now—Force B is *fake* force therefore no contact with Crook, but Force A is *real* and there is a battle contact in the Manassas Gap—Jackson has 10 Infantry Regiments, 3 Cavalry Squadrons and 4 guns whilst Getty, in a prepared defensive position, has 4 Infantry Regiments, 1 Cavalry Squadron and 2 guns.)

Getty immediately sends a courier to rest of his force at Ashby's Gap (14 miles south). Courier moves at rate of 5 miles per game-move but he draws a Chance Card which declares him to be wounded by guerillas causing a further 2 game-moves delay before he is capable of giving his message to General Wheaton. This means that the Manassas Gap battle will have been going on for 5 game-moves before Ashby's Gap force hears of it. (Day lasts 8 game-moves.)

General Wheaton's Chance Card only allowed him to send a third of his force (2 Cavalry Regiments and 1 gun) whilst remainder (6 Infantry Regiments, and 1 gun) followed two moves later, so that the cavalry force will not arrive until end 9th game-move, whilst second (Infantry) force, moving at 2½ miles per move, will not arrive until end 11th move.

At nightfall, end of 8th game-move, Jackson has not managed to destroy Getty's force, so he pulls out and moves south down

East side of Blue Ridge Mountains. Getty retreats towards Ashby's Gap by that road he has kept open. Losses are—Federal, 1 Cavalry Squadron, 1½ Infantry Regiments; Confederates, 1 Cavalry Squadron and 1 Infantry Regiment—under campaign attrition rules this is 25% of their actual losses. During his retreat, Getty meets Wheaton's cavalry reinforcements who also turn and retreat with him, later they meet the rest of Wheaton's force and Getty, who has panicked, carries on back to Winchester leaving Wheaton remaining in situ 5 miles from battle area with his 6 Infantry Regiments and 1 gun; he sends scouts forward.

Getty has also sent a courier to Crook when battle began, this rider does the 25 miles eastwards uneventfully in 5 game-moves. Crook draws Chance Card which entitles him to send off his 3 cavalry squadrons at forced-march rate (so that they will arrive at Manassas Gap at the end of 10th game-move, remainder of his force follows and are scheduled to arrive at the end of the 15th game-move.

After battle, in accordance with rules Jackson's force now splits again into a real and fake force—Force A moves south-west along Manassas Road towards Front Royal (actually being passed in the darkness 2½ miles away by Crook's forces moving towards Gap). Force B moves south-west down Eastern side of Blue Ridge Mountains, through Swift Run Gap and across to river west of Port Republic. In meantime, Force A arrives on river 2 miles north of Port Republic.

Sheridan's Army of Shenandoah has passed through Harrisonburg and on to Cross Keys, then to river where he finds Jackson's B Army with Port Republic immediately at its back. This B force is the real one and has been brought up to strength again by reinforcements from Staunton—so a contact has been made and battle is fixed. In meantime, Crook's force has reached Manassas Gap, found nothing and so the cavalry proceed east of Blue Ridge Mountains following trail of Jackson's B Army whilst he follows trail of Force A. Both of Crook's forces will arrive at Port Republic battle area 10 game-moves after start of battle (2 moves into the night).

Thus Jackson, even if he beats Sheridan, has to take on Crook's full strength force with his battle survivors, as there is not time for reinforcements to reach him from Staunton. This fact alone may well have decided Jackson to leave the Valley but he was beaten by Sheridan anyway! He now had Sheridan's

M

survivors, superior in strength to himself, chasing him through Port Republic with Crook coming up fast with another complete army. Owing to the fact that Sheridan had never lost contact with Jackson it was not considered fair at this stage to again split the Confederates into a fake and a real force.

So Jackson, realising the game was up, pulled out of the Valley, not having been very successful in his mission. Had he quickly beaten Getty in the Manassas Gap, split again into two forces and moved further up the Valley he would have had both Sheridan and Crook going the wrong way and only Wheaton's smaller force between him and Winchester and Harpers Ferry.

The reader may, initially, find this a bit complicated, but if he works out the move speeds, and realises that a day takes 8 game-moves, then it is not difficult to ally the movements of the forces on his map. It was a good campaign, lasted about three weeks and proved how hard it is for smaller forces to beat larger ones, except when they are strongly entrenched like Getty in the Manassas Gap!

# 22

# The Apache Uprising
## (American West 1870)

The Apache tribe consisting of 105 foot braves and 56 mounted warriors, live at the village of St Francis. They rise against the white settlers and soldiers and move southwards.

The 7th Cavalry Regiment and the 2nd Texas Infantry (formed in two squadrons of 15 men each and seven companies of 20 men each) are at Forts Lincoln and Grant. They move to fight the Apaches, but leave one company at each of the forts—they also have one 12 lb Napoleon gun and team.

Dotted around the countryside are four settlements—Riverside, Lake, Wooden Roof and Cedar Bend, each of which is defended by five settlers.

The campaign is a combination map/war-games affair and is decided on the following points system:

Each soldier killed counts 5 points to the Apache, each officer is 10 points and the Commanding Officer is 20 points. Each Indian killed counts 5 points to the soldiers. Map moves are made and battles take place in the usual way when map contacts are made, but fighting for the settlements takes place on the map (unless those settlements are reinforced by the soldiers when normal fighting takes place).

Each settlement destroyed counts certain points to the Apache—when the Indians reach a settlement a dice is thrown —1, 2, 3 there is a surprise and 4, 5, 6 the settlers have been warned. The settlement is then destroyed as under:

*With surprise element*—15 Indians destroy the settlement and settlers in two moves, win 25 points and lose two warriors (10 points).

20 Indians or more destroy settlement in one move, win 50 points and lose 3 warriors.

*Without surprise*—15 Indians take four moves to destroy settlement, win 25 points and lose 3 men.

20 or more Indians take two moves, win 35 points and lose 4 men.

If the Apache capture a fort they gain 100 points in addition

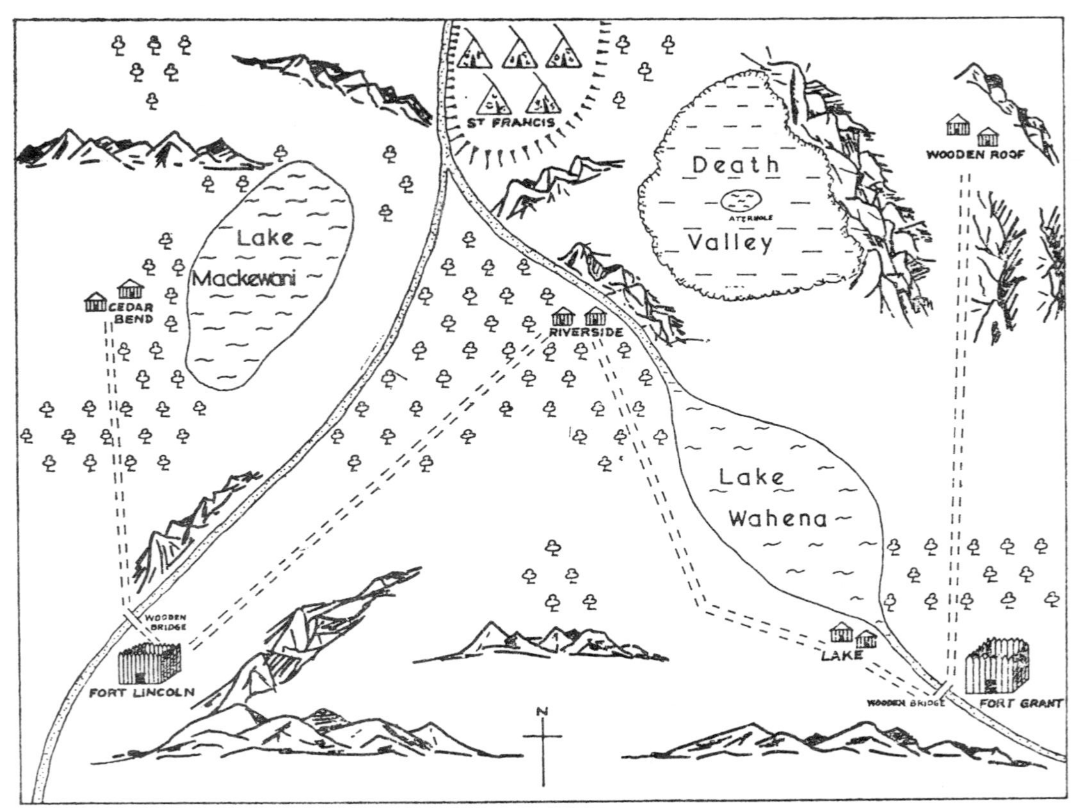

ST FRANCIS
Lake Mackewani
CEDAR BEND
Death Valley
ATERHOLE
WOODEN ROOF
RIVERSIDE
Lake Wahena ~
WOODEN BRIDGE
FORT LINCOLN
LAKE
WOODEN BRIDGE
FORT GRANT
N

to those scored through killing soldiers. If a supply waggon is destroyed, Indians win 25 points. Destruction of gun scores 50 points and may only be done if it is captured and 3, 4, 5 or 6 thrown.

The campaign is decided by the Indians gaining 500 points, or by the soldiers capturing St Francis village. If the Apache decide that they cannot possibly make this total they may sue for peace at any time.

### Map-moving

Soldiers move one square per move across country and two squares on roads, cavalry double. Indians move $1\frac{1}{2}$ squares across country and 2 on roads, cavalry 4 squares. The lakes cannot be crossed by soldiers but may be crossed by Apache, who take one move to make birchbark canoes and embark, rivers take one complete move to cross for everyone. Canoes move 2 squares per move. Death Valley is crossed at half speed and one move must be spent at waterhole, if force are denied waterhole they lose 25% of their strength for each move they are denied water.

### Table-tactics

Usual moves for soldiers, Apache foot move 9 in. at all times. Firing by volleys as usual, but Apache may fire fire-arrows at half usual rate. Fire-arrows can destroy waggons—throw 4, 5, 6 to denote a hit and then three hits. Same applies to fort walls which may have a breach burned in them with three hits.

If Apache attack soldiers in road running through woods or bluffs, etc., they may try to secure surprise—if they throw 4, 5, 6 this is achieved and they have first move and first firing when within range.

After battle, throw a dice for each man killed—all 1's and 2's are permanently out of campaign, remainder may count for future battles.

# A 'Potted' Campaign
## (Franco-Prussian War 1871)

This small Franco-Prussian war affair is a 'potted' campaign of a simple nature that can be fought between two players without the services of a third party as umpire. It has the merits of being capable of completion in a single evening or at the most in two sessions.

A map was drawn and squared 6 by 6 to give a total grid of 36 squares each one of which represented one of the 36 drawers in the matchbox 'chest'. The initial forces at disposal of both commanders are as follows:

|                 | *French* | *Prussian* |
|-----------------|----------|------------|
| Light Infantry  | 2        | 2          |
| Line Infantry   | 8        | 7          |
| Light Cavalry   | 1        | 1          |
| Heavy Cavalry   | 1        | 1          |
| Horse Guns      | 0        | 3          |
| Heavy Guns      | 2        | 2          |
| Field Guns      | 1        | 0          |
| Howitzer        | 1        | 1          |
| Mitrailleuse    | 2        | 1          |
| Pontoon Section | 1        | 1          |

The campaign opens with the French force in position guarding the railway bridge at the southern part of square 'P' and the river bridge at the northern part of square 'V'. The entire German force are positioned in square 'W'. The Prussians plan to force river crossings and have the choice of four alternative methods.

1. By rail and road bridges in squares 'P' and 'V'.
2. By road bridge that overlaps squares 'Z' and 'FF'.
3. By road bridge (partly destroyed) in square 'F'.
4. By pontoon bridge to be erected at any selected point on the river (this takes one map-move or three table-moves to construct).

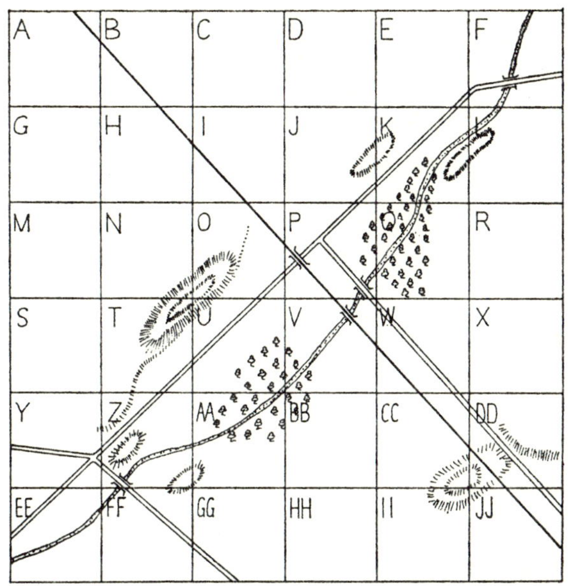

A B C D E F
G H I J K L
M N O P Q R
S T U V W X
Y Z AA BB CC DD
EE FF GG HH II JJ


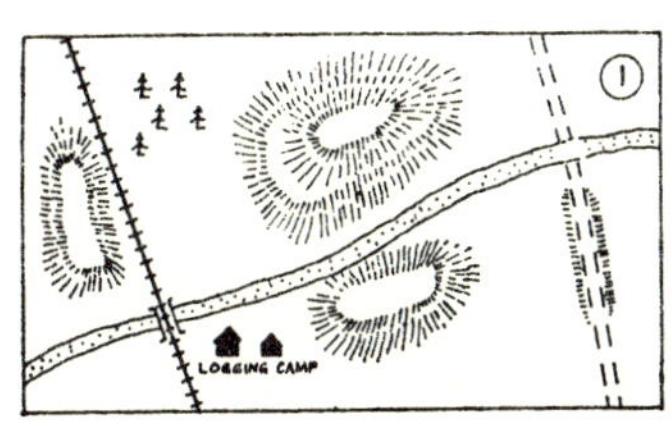

1
LOGGING CAMP


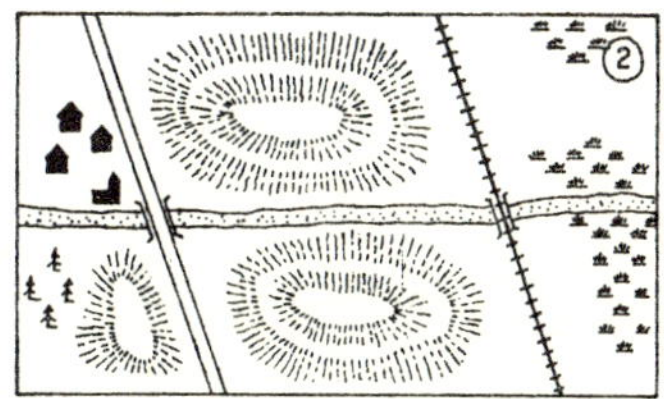

2


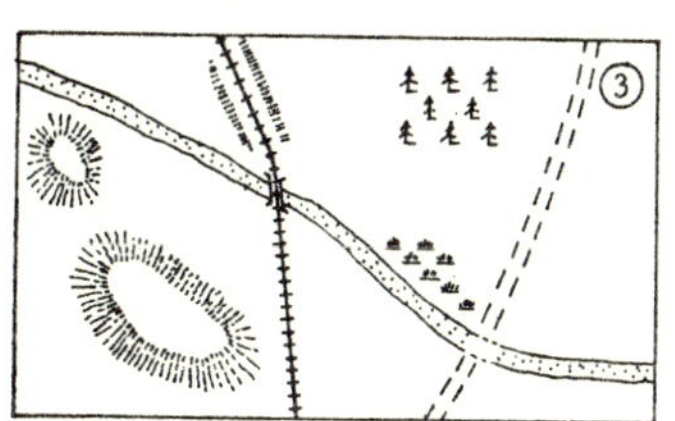

3

*Method*

At nightfall on the 9th January, 1871, the Prussians begin to move all their forces (collectively or individually as desired). They are represented by numbered discs all of which commence in the box 'W'. The Prussians move first and they transfer their counters one square (box) in any direction.

The French now move one square.

Whenever French and Prussian discs are found in the same box their points value (with 20% optional value) is declared if they are physically able to be in contact. Both sides have to agree on a contact before any battle can take place. When such a contact occurs and is agreed by both sides then the specific terrain is laid and forces assembled.

History shows that few armies seem to go into war with accurate maps and this campaign is no exception. The overall map on which the moving takes place contains only the barest topographical features, i.e. the road, river, railway and hills and some forest. Obviously 'nearer to the ground' there are other topographical features which can be put in and will make a battle more interesting to say nothing of perhaps making it more difficult for one or other of the commanders. To carry out this system, three separate Terrain maps (numbers 1, 2 and 3) have been drawn, to include not only additional features but also to be slightly different to the Main map. When a contact occurs in one of the principal areas then the resulting battle is fought on the relevant Terrain map. The rail and road bridges in squares 'P' and 'V' are covered by Terrain map number 2.

The rail bridge in squares 'Z' and 'FF' is covered by the Terrain map number 1 whilst the rail bridge in square 'F' is covered by Terrain map number 3.

In the event of a pontoon bridge being laid at some other point in the river or the action taking place in an area that is obviously outside those covered by the Terrain maps then a further map or maps will be compiled by mutual agreement between the two contestants.

When a contact is made and battle joined, any troops outside the area of the Terrain map may move to that area on the Main map at the rate of 3 moves on the war-games table equals 1 square on the map.

A few 'local' rules were added before the campaign actually commenced:

1. Rail bridges were merely open plank erections with rails laid on top. This rendered them unfit for cavalry and artillery.
2. The rail bridge in square 'F' was partly destroyed and could only be crossed by two infantry units per map-move.
3. At the fords (shown only on Terrain maps) all arms could cross at the rate of a complete force in one map-move or three units (any arms) per game-move.

Pontoon bridges could be laid as discussed earlier and crossed by three units (any arms) per game-move.

Such a map, with its collection of smaller Terrain maps and its general method of application, can be used for more or less any part of the Horse and Musket period or even up to modern times. Such a set of maps is always valuable for future use and should be retained after serving its initial purpose, probably to be sorted out and used again in later years when another 'potted' campaign is required to fill in an evening or a short war-gaming period.

# 24

# The Boer Revolt
(South Africa 1883)

This stimulating Colonial campaign sowed the seeds that later led to the vast panorama of operations in East Asia. Originally fought out between two players, it can conveniently include a third man to handle the surprise factors or it can become a Club Project with British forces, Commandos, natives and the gun-boat allocated to individual members.

*Narrative*
Off the map and far to the North a British Cavalry patrol sight a Boer Commando moving southwards. At a point immediately north of the Camp (but still off the map) the patrol survey the Boers' movements without being detected. They estimate the strength of the force to be between 150 to 200 men formed of horse and foot with waggons and artillery. Realising that this is a hostile force, the patrol commander turns south to warn of the impending danger. On re-entering the map, the patrol split up as they proceed to the various settlements and towns that spread out before them. Near the Camp a Boer scout is captured—he reveals that they have only seen part of an invading force and that there are two other similar-sized Commandos on the move at other points. In the course of conversation he mentions that all is not well with the Commando leaders who are jealous of each other so that it is extremely likely that they will act separately and only join forces in an extreme emergency (see Technical Notes for further details on this point). The cavalry patrol continues on their warning journey.

*Technical Notes*
The main map used for this campaign was scaled ½ in. to 12 in. on the war-games table, covering an area equal to 16 war-games tables (8 ft. by 5 ft.) placed 4 by 4. The movement rates on the map are as follows:

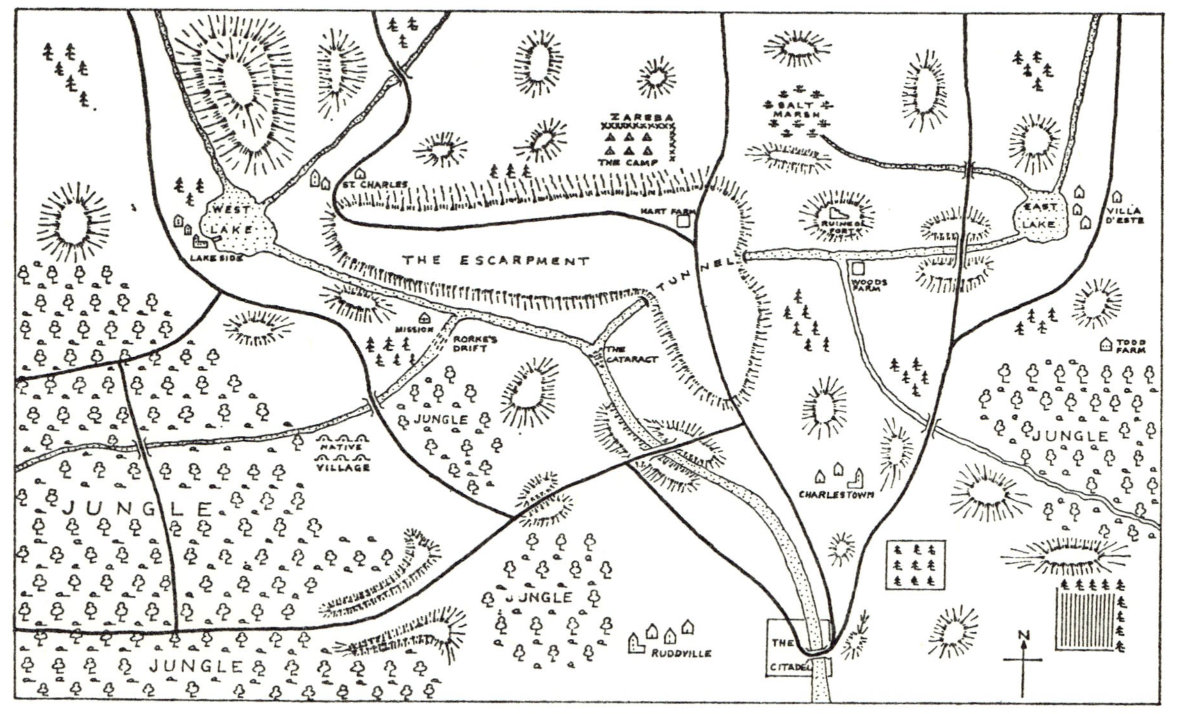

ZAREBA
THE CAMP
SALT MARSH
ST CHARLES
WEST LAKE
HART FARM
RUINED FORT
EAST LAKE
VILLA D'ESTE
LAKE SIDE
THE ESCARPMENT
TUNNEL
WOODS FARM
MISSION
RORKE'S DRIFT
THE CATARACT
TODD FARM
JUNGLE
NATIVE VILLAGE
CHARLESTOWN
JUNGLE
JUNGLE
JNGLE
RUDDVILLE
THE CITADEL
N
JUNGLE

|          |               |          | *On table* |
|----------|---------------|----------|------------|
| *Infantry* | Cross country | $\frac{1}{2}$ in. | 12 in. |
|          | On road       | $\frac{3}{4}$ in. | 18 in. |
|          | In jungle     | $\frac{1}{4}$ in. | 6 in. |
| *Cavalry* | Cross country | $\frac{3}{4}$ in. | 18 in. |
|          | On road       | 1 in. | 24 in. |

The patrol may add 50% to these rates whilst carrying out their warning duties.

|          |               |          |        |
|----------|---------------|----------|--------|
| *Boers* | Cross country | $\frac{3}{4}$ in. | 18 in. |
|          | On road       | 1 in. | 24 in. |
| *Civilians* | Cross country | $\frac{1}{4}$ in. | 6 in. |
| (and | On road | $\frac{1}{2}$ in. | 12 in. |
| livestock) | | | |
| *Gunboat* | Up-river | $\frac{3}{4}$ in. | 18 in. |
|          | Down-river | $1\frac{1}{2}$ in. | 36 in. |
| *Waggons* | Cross country | $\frac{1}{4}$ in. | 6 in. |
|          | On road | $\frac{1}{2}$ in. | 12 in. |
| *Guns* | Cross country | $\frac{1}{4}$ in. | 6 in. |
|          | On road | $\frac{1}{2}$ in. | 12 in. |

This was one of those affairs in which the operations are so planned as to allow a relatively small number of figures to be used to represent a number of forces used in separate operations. Thus the reason for the apparent jealousy between the leaders of the Boer Commandos was due to the fact that there were only 150 to 200 Boers available in all but as they were only likely to fight as a separate Commando this worked out well.

The essential point at these opening stages of the campaign was that no British force should take any action or make any movement until they received warning from the cavalry. Then each local commander was free to open his Emergency Orders provided for just such a contingency as this. These orders gave a considerable amount of discretion to each commander, allowing him to decide whether to defend his immediate area or to fall back or even to take the offensive. Each set of orders concluded with the following paragraph:

'*All* civilians and their livestock in outlying farms and in the towns and villages will be evacuated under adequate escort to places of safety. Should this not be possible, these villages and farms will be defended by bodies of troops specially detailed for that purpose. Under no circumstances will civilians or livestock be allowed to fall into the hands of the enemy.'

It is worth pointing out that at no time could civilians and livestock move at a speed beyond their normal slow rate, thus limiting their escorting military forces to these same rates of movement.

Throughout the campaign all forces were numbered on a points basis—Infantry equalling 1 point each, Cavalry equalling 2 points and guns being worth 10 points.

*Disposition of British Forces in the area*

| | |
|---|---|
| Citadel | 150 points |
| Charlestown | 50 points |
| Ruddville | 100 points |
| Lake Side | 100 points |
| St Charles | 100 points |
| Villa D'Este | 100 points |
| The Camp | 60 points |
| Force on road near Hart Farm marching to Camp | 60 points (at Tunnel at start) |

Gunboat (at Citadel), Naval detachment of 30 men and Gatling gun. A total of 400 infantry, 120 cavalry, 12 guns plus gunboat crew. Civilians and livestock situated at:

Todd Farm
Woods Farm
Hart Farm
Mission at Rorkes Drift

At its base at the Citadel, the gunboat may move up-river. In this particular campaign, by chance the gunboat set off on a routine patrol northward on the same day as the Boers were sighted. At the junction of the arms of the river are the cataracts which take one full move to negotiate. To the north-east the river passes through the escarpment by a huge natural tunnel and the captured Boer has disclosed that one of the Commandos has with it a waggon-load of explosives to blow up the entrance to the tunnel through which the river runs. Should this be successful, the gunboat will be unable to pass through the tunnel and will be trapped either to the east of the tunnel or else will only be able to patrol west of the tunnel (with only two players taking part in the original campaign both knew of the existence of this waggon-load of explosives. With the inclusion of an umpire or if the campaign is being fought as a Club Project then the British need not know of this threat. Similarly, far greater elements of surprise can be introduced by the umpire, for example the Boers need not know that they have been sighted nor that their Scout has been captured).

To blow up the tunnel, the Boers must get their waggon over one of the tunnel mouths where it must remain for two complete game-moves, then two dice are thrown and 8 or over is needed to blow in the tunnel mouth. They have no second chance of doing this.

Just as the civilians, their cows and sheep, etc., made themselves a darned nuisance on the war-games table so they are making their noisy presence felt in these pages because it has not been mentioned that these civilians must be guarded by a military force equal to at least half the numerical strength of the civilians. In the early part of the game the war-gamer taking the British showed a marked and callous disregard for the welfare of the civilians, hastening his troops into the zone of operations and leaving the unfortunate settlers to fend for themselves. Hastily a rule was compiled which stated that any individual civilian or piece of livestock that fell into the hands of the enemy or were killed through non-compliance with the original ruling meant that the British strength at the Citadel (150 points) is reduced by 1 point per lost civilian, cow, sheep, ox, ass, etc., etc. The war-gamer commanding the British felt so strongly about his obligations towards the 'damned civilians' that special care was taken when compiling rules for the East Asia campaign so that the settlers were transformed from meek, mild and panic-stricken farmers into tough characters strangely resembling the cowboys of the Wild West. Their exploits during the German attack on Kerka and Fort Windsor have become Frontier legends!

If one thing stands out in recounting table-top campaigns it is that, just as in real-life warfare, what occurs on the battlefield is merely the short culmination of far lengthier preliminaries. It happened during the South African uprising of 1883 and the point is made here as an explanation (or an apology) for even more prolonged technical details. For example, how was the gunboat made to chuff and puff happily up and down the tortuous rivers?

Well, its rates of movements have already been given and it

is known that it has a crew of 30 men with a Gatling gun (or 35 men without a Gatling gun). The boat is armed with a 6 pdr. gun fired to normal artillery rules. The gunboat is always considered to be in the centre of the river unless specifically designated as being tied to the bank or at a wharf. This means that its gun will never fire under 12 in. range and that fire against the gunboat from the bank of the river will also be at 12 in. range. If the gunboat so desires it can come in closer to the bank and reduce its firing range to under 12 in. but similarly it must receive fire at this decreased range.

The gunboat can tow four boats each capable of holding 5 men—when performing this task the little steamer will only move at half rate. To embark or disembark men takes 3 in. The Gatling gun can disembark and open fire or it may disembark and move in the same move, but not both. The Gatling gun is pulled by four sailors and moves at the rate of 6 in. per game-move on the table or $\frac{1}{4}$ in. on the map.

During the actual fighting of the campaign the gunboat set off on its routine on the same day as the cavalry patrols saw the Boers, it had got to the cataract and was slowly navigating it when the patrol began their warning ride. We already know that the cataract takes one full move to negotiate and it is also worth mentioning that the gunboat may only turn at the lakes or at the Citadel. To avoid a heated over-the-table argument it is worth a rule that the gunboat can proceed at half-rate in reverse!

The Boer artillery is capable of knocking out the gunboat. Rules to choice can govern this factor—in the campaign under description the following methods were used:

Boer guns firing on the gunboat at ranges over 24 in. needed to score a 6 on the dice to hit.

Guns firing on the boat at ranges over 12 in. needed a dice score of 5 or 6 to hit.

Guns firing on the boat at ranges under 12 in. needed a 4, 5 or 6 for a hit.

Once a hit is scored, two dice are thrown and a total of 12 sinks the gunboat and kills half the crew. Two guns firing on the gunboat must throw four dice and score 21 to achieve the same result. Three guns firing on the gunboat need to throw six dice and score 29. If one less is totalled by any of these combinations then the boat remains motionless for two moves; 2 less mean it stands still for one game-move. Scores below that could kill crew members by normal artillery scoring methods.

This campaign involved only the forces specified at its commencement so that some form of attrition had to be devised. In this particular campaign all casualties were totalled up at the conclusion of the battle and 20% of the points total was eliminated for good. An alternative method is to throw a dice for each individual casualty at the conclusion of the battle—5 or 6 means that the man is slightly wounded and returns immediately to duty; 3 or 4 means that he is wounded and will miss one battle, returning for the battle after that, whilst 1 or 2 means that the man is killed and permanently out of action. By using a War Diary (explained elsewhere in this book) specified disability periods can be nominated for individuals or groups of casualties.

War-gamers may consider that what with British forces, Boers, farmers and their livestock, they have enough on their plate but spice is added to the campaign by allowing those troublesome natives who infest the jungles to the south and south-east of the map to take advantage of the British preoccupation with the invasion and rise up. The manner in which this was carried out during the campaign forms a separate little narrative.

On the conclusion of the 8th move (whether map-move or game-move is left to the choice of the participants) a cavalry patrol were considered to be moving westwards along the Lower Jungle Road towards the off-map village of Umbala where they had to arrest a native suspected of sacrificial murder. Disturbed by this apparent invasion of their territory, the natives in the jungle area concerned decided to resist the patrol. Now arises the need for a decision—are the patrol to be considered expendable or should they be rescued by a relief column from the Citadel? Or is the Commander-in-Chief at the Citadel justified in reserving his entire strength to fend off the very strong Boer invasion? Inflamed by the incessant beat of war drums, the natives residing in all the jungles to the south and south-east of the map decided to rise and destroy the white men who had taken over their country. In the original campaign this decision was determined by drawing from a number of slips of paper, some bearing a time and date whilst others were blank. The tribes drawing blank slips did not rise up whilst the others surged forth at the duly appointed time, on foot and by canoe.

Although it would seem that the objectives of this game were clear, i.e. either the British threw back the Boers or the Boers cleared the British out of the area, it was deemed necessary for reasons beyond recall to devise a system of awarding points to determine the campaign winners. A definite victory in a battle secured 25 points plus 15 points if the other side withdrew whilst a drawn battle gave 10 points to each side. One point was awarded for every enemy infantryman killed; 2 points for every enemy cavalryman killed and 10 points awarded for every enemy gun destroyed or captured.

### The course of the campaign

There were five roads running from north to south and it was possible for the Boer commandos to move southwards down any three out of the five roads that ran down from the north of the map. Because of their waggons and guns plus the fact that the river was only crossable at the bridges it was thought unlikely that the Commandos would move across country. In the event, number one Commando came in on the St Charles road; number two Commando moved east of the Camp and Hart Farm down the road that passed across the roof of the tunnel whilst number three Commando came in on the most easterly road to attack Villa D'Este.

Number one Commando attacked the town of St Charles garrisoned by a force of 60 infantry, a squadron (15 men) of cavalry and a gun. Warned in time by the cavalry patrol, this force had used their time well to fortify their position so that they administered such severe casualties to the attackers in the early stages of the battle that the Boers retired to lick their wounds and virtually took no further part in the campaign. This action confirmed the report of the captured Boer scout that the Boer commanders had little intention of collaborating with each other because, in the light of future events, it seemed that number one Commando could have swayed the entire course of the campaign had they taken St Charles and then moved eastwards along the escarpment road to join with number two Commando in the area of the tunnel.

Finding it a grave handicap to marshal the civilians and their livestock back into safety, the British forces were only able to group in sufficient strength to fight back when number two Boer Commando had reached the middle of the map on the road immediately south of the tunnel. Here they came up against a British force of approximately equal strength to themselves. In

N

the hard battle that followed, although the British had heavier losses than the Boers, they managed to do enough to win the battle so that the Boers fell back.

Number three Boer Commando moved down the road furthest east on the map and attacked a surprised Villa D'Este. As might be reasonably expected the British were defeated but they put up a brave fight, managing to save about 30 infantry who fell back down the road. Before leaving, the garrison damaged their gun beyond repair. A noteworthy point of this battle was the manner in which the civilians rallied round and fought alongside the troops, losing 75% of their number in the action. This affair lasted for six game-moves and in the resulting tot-up, the British lost 60 points for killed soldiers, 10 points for a destroyed gun and 20 points for a civilians, making a total of 90 lost points. The Boers lost 20 points. The final result was that the British had gained 20 points but lost 90 giving them a total of minus 70 points, whilst the Boers had lost 20 points but gained 90 giving them a positive total of 70 points.

Whilst the battle south of Tunnel Ridge had been going on, the Boers had made their attempt to blow in the mouth of the tunnel but soon discovered that all their available explosives could make little impression upon the vast rocky gorge that formed the entrance to this amazing natural tunnel that passed under the huge escarpment. All these events had taken up 14 game-moves when the position was as follows:

A British force held St Charles whilst a sullen Boer Commando stood off and licked their wounds.

Number two Boer Commando were concentrated at the crossroads near the Hart Farm facing a British force in a good natural position stretching from east to west across Tunnel Ridge.

Number three Boer Commando, having taken Villa D'Este, had moved southwards down the road to reach the bridge cross-in, the river on the Charlestown Road.

Now followed the crux of the campaign as the British commander, with all the skill of a bookmaker laying-off his bets, juggled around his forces to defeat the invader. Conscious that the garrison of St Charles was unnecessary as number one Commando, even if inclined, could not proceed southwards except by way of the road crossing Tunnel Ridge, the British commander decided to use them elsewhere to better advantage. At the same time he had to counter the threat to his right rear and indeed to the Citadel posed by number three Commando coming down from Villa D'Este. Towing a string of rafts behind it the gunboat moved up-river and evacuated civilians and livestock from St Charles, disembarking them at the Mission at Rorkes Drift. Next, the commander had to reinforce the small force that had retreated from Villa D'Este and were holding the bridge on the Charlestown Road. A cavalry force with artillery and half his infantry were detached from the force that lay across Tunnel Ridge and moved south-east towards the bridge. At the same time, the commander allowed the Boers to be aware of this apparent weakening of his holding force and, so encouraged, they took the bait and attacked. The gunboat immediately moved to the western entrance of the tunnel where it disembarked its sailors and marines and then remained as a most effective gun position that gave trouble to the Boer right flank throughout the battle that followed. Unknown to either number one or number two Commando the garrison of St Charles stealthily abandoned the town and moved eastwards along the escarpment road, coming in on the rear of the Boer force when they were hotly engaged in their attempt to force the British position across the ridge.

Demoralised by the arrival of an enemy force when they had expected reinforcements in the shape of their own number one Commando, the Boers broke on all sides and the battle was over. Probably, had number one Commando realised what was going on and followed the St Charles garrison down the road things might have been different, but in the event number one Commando continued their passive policy and on discovering that number two Commando were routed, disappeared back to their farms and villages.

Down at the bridge on the Charlestown Road the almost full strength Commando was initially held off by the very much smaller British force, who were reinforced in the middle of the second game-move by the British cavalry and artillery from Tunnel Ridge. At the middle of the third game-move the British infantry units arrived and, although still less in numbers than the Boers, the British force were in positions of sufficient strength to be able to withstand all assaults. At nightfall both sides had suffered approximately the same casualties, but during the night the Boer Commando withdrew and the invasion had ended. It later transpired that they had discovered by some means of the disastrous fate of numbers one and two Commandos which caused them to realise that even success at the bridge would not achieve the objectives of the invasion.

Of course the victorious British forces were not able to completely leave their positions and flood back to cope with the native rising which, to everyone's surprise, was blazing up like a fire made of wet wood. A small native force ambushed the cavalry patrol in the rocky pass through which the southern jungle road ran but the mounted men were able to repel and out-distance the natives, killing many of them for little loss themselves. The tribesmen of the five or six jungle sections that lay on either side of the river and those from the village were engaged in a bitter dispute as to who should lead the rising and to which tribe should fall the glory of destroying the white men. So fierce did these arguments become that fighting actually broke out between two of the tribes so that neither were able to take any effective part in the sporadic fighting that followed. The Mission at Rorkes Drift was attacked and gallantly defended by civilians who were relieved when the gunboat arrived and landed its sailors and marines who, with their spitting Gatling, put the natives to rout. After this, the gunboat made its way down the river to shell and destroy the native village. At this point an infantry force who had marched from Tunnel Ridge met the gunboat and were towed behind it on a train of log rafts. This force arrived at the westmost bridge, landed, and in a pitched battle proved that native courage and spears were no match for discipline and firearms. This just about wrapped up the native rising and the whole area settled down to a period of peace and prosperity—before someone decided that it was a convenient and suitable area for further operations!

# 25

## The Relief of Fort Mazzoni
### (Mythical Franco–Prussian Colonial Campaign 1885)

The Germans—510 points—are besieging the French (170 points) in Fort Mazzoni.

A French relieving force—340 points—leave Roussell and have to relieve Mazzoni in 10 days. It takes 1 day to cross one

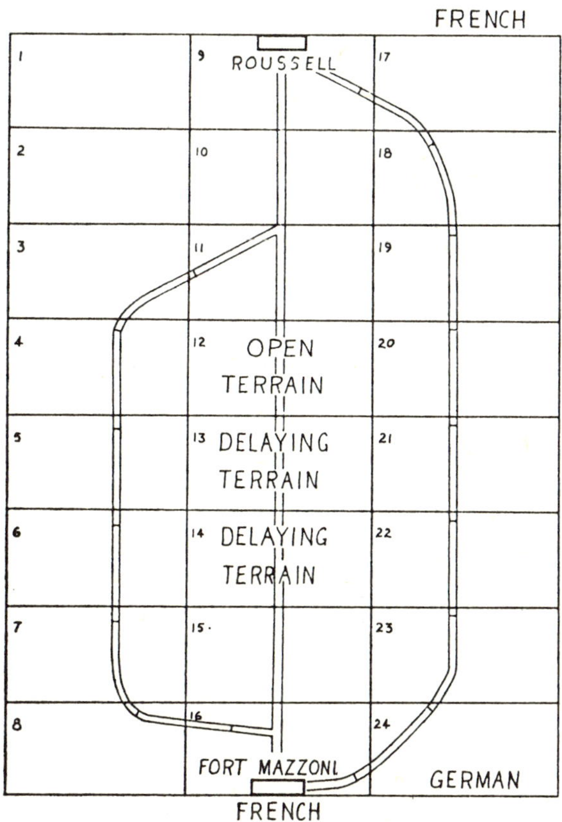

rectangle of the map. Cavalry moving by themselves do $1\frac{1}{4}$ squares per day.

The German commander detaches $\frac{2}{3}$ of his force (340 points) who march north to confront the French. This German force is the same strength as the French but differently composed— part of it is 1 Regiment of Lancers who may either be used with main body *or* detached to watch either of the East and West roads.

Only the centre road is capable of taking entire forces of this size at one time—so it is certain that this road will be used by the French.

The Germans may either mass and fight their entire 340 points in the open terrain or they may split into two equal halves and fight delaying actions to prevent the French arriving in their required 10 days. In these delaying areas the Germans will have time to fortify. In the event of a German defeat in the 'open' area, they may re-group and fight again in the 'Delaying' area (square 14).

A selection of five Terrain maps are given to the German commander. Two of these maps will be suitable for the 'Open' terrain (one of which may be selected) and he may pick two of the three other maps for his 'Delaying Areas'. They will then be laid out, when required, on the table.

The French force in Fort Mazzoni may endeavour to break out to aid their relieving force.

The duration of a battle can be decided by agreeing that a day lasts *eight* game-moves. By counting the number of moves over which the battle is fought, that portion of the day that has passed may be estimated.

# 26

# A Colonial Campaign
## (German South West Africa 1916)

*Narrative*

German Colonial forces under General V. Vorbeck are being pursued by superior British forces under General Allenby. The Germans are formed in three columns and the British in four. The country over which they are operating is mixed and consists of many varieties of terrain; it only has one road running from north to south and, dotted over it, are five waterholes. To survive it is essential that each column tries to obtain the waterhole in its area. Thus, the waterholes and the road are the objectives and the side that does not possess them will not last long enough to have more than perhaps one second attempt.

The country is unmapped and unknown to the opposing forces; the exception to this is that each side has *one* guide who can only accompany *one* of the columns. But even this guide does not know too much about the country—although he is aware of the location of the waterhole in his own particular area.

Each column has a baggage-train accompanying it; when contact is made with the enemy and a battle takes place, this baggage train has to be established at a point marked on an 'off-table' map of the area immediately behind the battlefield. This area is not known to the enemy unless they have a guide with them who will be able to tell them of **its** peculiarities and features so that they can possibly work out where the baggage might be located. The campaign was fought to the '1917' Rules Handbook obtainable through *War-gamer's Newsletter*.

*Technical notes*

This is a three-cornered affair and requires a third man to work the map factors. Both the generals and the third man have talc-covered sheets $12\frac{1}{2}$ in. by $12\frac{1}{2}$ in. divided into $2\frac{1}{2}$ in. squares—thus each sheet has 25 squares in 5 columns, each column being painted a different colour. Initially, each square is blank.

The third man will mark in on his sheet the road running from top to bottom (N to S). He will not reveal its course but will tell each general separately the square in which it begins at his end of the map. The third man then draws a 'terrain-card' for each square of his map—these cards bear merely the name of the type of terrain (thus, 'scrubland' or 'undulating open plain' or 'hilly', etc.). In turn he draws a card for each square and writes in that square the type of terrain it is.

The British general then selects four of the five terrain-columns in which he intends to move his forces—the third man tells him what sort of terrain the first square of each column consists of. The German general then selects three terrain-columns for his three forces and he is also told what sort of terrain is in the opening squares of his three columns. Then each general is given the type of terrain in the *next* square forward for the force that has the guide with it—the guide knowing what lies ahead to that extent.

The third man now marks in the waterholes on his map—there is one to each of the five terrain-columns and it can be in one of the three squares in the middle of each terrain-column (thus it would be numbered 2, 3 or 4 running downwards). He does not reveal this to either general except to the force that has the guide with it for their terrain-column.

The British and German forces are divided up into columns as detailed elsewhere in this plan; they are roughly of the same strength although not necessarily of the same composition. A force may move as a whole—in which case it moves at the speed of the slowest arm in it—or it may detach faster elements and divide itself into two—again each half moves at the speed of the slowest part.

*Movement rates :* See attached chart.

*Method of movement :* Each force has a 'Movement Card' on which its order of march is detailed (it arrives on a battlefield in that strict order). The general gives each force's card in turn to the third man—telling him which column of terrain they are using—and also marked with the movement rate of the force or forces so that the third man can mark his map accordingly.

When the third man notes that two enemy forces have met in any one square he announces that a contact has been made—either general may refuse the contact and withdraw back to the previous square. When there are already troops of one side in a square this prevents any forward movement by the other side—each side announcing whether or not they wish to engage and the side refusing withdrawing back to the previous square.

A unit or force may fortify a position if it stays put in that square without moving for *one* move.

As the British have one more force than the Germans they will be moving in a terrain-column that has no German force therein. This British force will not know this until it has traversed the entire terrain-column and arrived at the far end. Then the force may move at specified rates in any direction to engage enemy forces. But no British or German force may be reinforced to be bigger than it was originally—in other words, all forces in each terrain-column never blend but work separately throughout the campaign.

*Attrition :* At the conclusion of a battle, each General throws a dice for every man or piece of equipment lost in that battle. For every man or item that gets a 1 or 2 that item is permanently lost; otherwise it can be returned to the force from which it came.

*Defeats :* A defeated force may have *one* chance only to give battle again (with its survivors after the dice attrition throws). It may fall back to the preceding square or any square behind that and stand if pursued; if not pursued it waits for 1 move and then attacks. But no force can last longer than 3 moves after a battle without possessing a waterhole—a battle is considered to take 1 move.

*Terrain :* A=Scrubland.  
       B=Undulating open plainland.  
       C=Hilly country.  
       D=Rocky country.  
       E=Marshy country.  
       F=Mountainous country.  
       G=Wooded country.

*Infantry* move    1 in. per map-move through C, E, F+ and G.  
             $1\frac{1}{2}$ in. ,, ,, ,, ,, A, B and D.

*Cavalry* move    $1\frac{1}{2}$ in. per map-move through C, D, E, F+  
                     and G+.  
            2 in. ,, ,, ,, ,, A and B.

Camels  
Mule Gun Team } as for cavalry.

*Horsed Gun Teams* move    1 in through G+.  
                    $1\frac{1}{2}$ in. ,, A, C, D, E+ and  
                              F+.  
                    2 in. ,, B.

*Lorries and Staff Cars* move 1 in. through D and G+ and E+.
              1½ in.    ,,    F+ and A.
              2 in.    ,,    C.
              2½ in.    ,,    B.

*Armoured Cars* move 1 in. through D, E+ and G+.
              1½ in.    ,,    A and F+.
              2 in.    ,,    C.
              2½ in.    ,,    B.

*Light Tanks* move 1 in. through D, E+, F+ and G+.
              2 in.    ,,    A and C.
               2½ in.    ,,    B.

*M/Cycles* move 1 in. through E+ and G+.
              1½ in.    ,,    A, D and F+.
              2 in.    ,,    C.
              1½ in.    ,,    B.

*Tractor-drawn Gun*—as for infantry.

+ means that movement is only possible on specified tracks.

All movement is *doubled* on the road (this does not mean tracks).

*On Table:* 1 in.=6 in.
           1½ in.=9 in.
           2 in.=12 in.
           2½ in.=15 in.

*Baggage Trains*—a force losing its baggage train is completely defeated and out of the campaign unless it can recapture it next move or capture the enemy's!

# 27

# A Punitive Expedition to the Pushna Valley
## (North West Frontier of India 1936)

For more than a hundred years the North West Frontier of India was a constantly running sore in the side of British rule. Nevertheless, at the same time it provided an ideal battle-training ground over which generations of British soldiers of each and all of the County regiments learned the basic facts of battlefield survival. The enemy were wily tribesmen who knew their hill country like the back of their hands so that they could blend into the rocky background and disappear or reappear at will. They were crack shots, frequently using British rifles that they had stolen with all the skill of practised cat burglars. Against such merciless and practised enemies, the British soldier had to learn to take cover correctly, to patrol, to fire accurately and in a controlled way—all factors that later stood him in good stead in major wars.

There was no period from the early years of the 19th century up to the end of British rule in the late 1940's when there was not some small British force engaged with troublesome tribesmen in the rocky gorges of the Frontier. In later years, some of the 'fun' departed from these operations when aircraft destroyed the almost delicate tournament-like atmosphere of these engagements. Of course, machine-guns, armoured cars and even light tanks had run their part in altering the complexion of the skirmishes but the wily tribesmen had, in their own inimitable way, managed to discover means of combating these modern weapons so that the balance had not been unduly affected. In the small campaign that follows, these factors are reflected and as a matter of interest it may be revealed that in the two versions that have been fought, the British achieved their objective in one but were completely defeated in the other.

Elsewhere in this book has been mentioned the merits of a battle or campaign that possesses a narrative to give it background interest. This punitive expedition has a recognisable objective, it is an operation carried out in a reasonably orthodox manner and it realistically simulates the type of Frontier affair that might have taken place during the 1930's.

It began on the 1st November, 1936, when the officer commanding A Company of the 1st Battalion the Hampshire Regiment stationed at the Razmak Cantonment received the following message:

*Subject: Punitive Expedition against the Malik-Ghazis of the Pushna Valley.*
From: The Officer Commanding 1st Bn. The Hampshire Regiment.
To: Major J. B. Hollands, M.C., Commanding A Company, 1st Bn. The Hampshire Regiment.

At first light on 5 November, 1936, you will command troops as detailed below in a punitive expedition against the Malik-Ghazis of the Pushna Valley. The object of the expedition is to proceed into the valley quelling any opposition and to destroy the mud-fort of the Malik-Ghazi tribe. On obtaining the objective, the force will return to Razmak Cantonment.

*Own Troops*
A Company consisting of:
*Headquarters Platoon*—Officer Commanding.
Lieutenant.
Sergeant.
2 Light Machine-gun teams.
8 Riflemen.

No. 1 Platoon ⎫
No. 2 Platoon ⎬ Each Platoon to consist of Lieutenant
No. 3 Platoon ⎭ Commanding; 1 sergeant and 12 riflemen.

*Total strength of force:*
5 Officers.
4 Sergeants.
44 Riflemen.
2 Light Machine-gun teams of 4 men each.
This force will be conveyed in four lorries.

The force will be supported by two light tanks and an armoured car of the 3rd Light Company of the Royal Tank Regiment.

*Enemy Troops*
Believed to number about 160, about a third of this force may

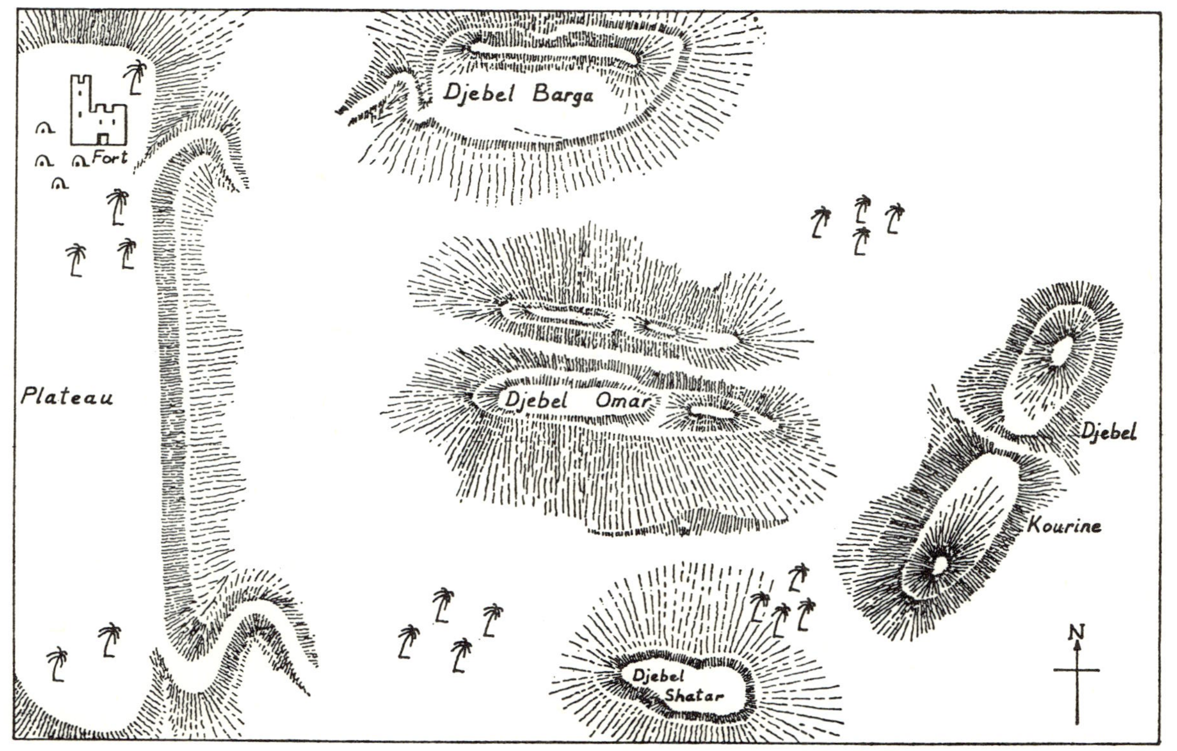

Fort
Plateau
Djebel Barga
Djebel Omar
Djebel Shatar
Djebel
Kourine
N

be camel- or horse-mounted. They are armed with rifles and a plentiful supply of ammunition. Information has been received that home-made land mines have been laid in the valley and its approaches and that the tribe are familiar with the use of petrol bombs.

*Casualties*
A First-Aid station will be set up in a safe position immediately on arrival in the valley. It is essential that no wounded man be left behind to the mercies of the tribesmen and each man who is wounded will be taken back to the First-Aid station or conducted back to safety by a comrade. A lorry may be detailed to accompany the troops as they move forward to pick up the wounded and return them to the First-Aid station.

*Destruction of Fort*
Explosives will be carried on by a squad detailed for the purpose or in a lorry that will accompany the force.
30 Oct. 1936.                         T. Camberley,
                          Lieutenant Colonel Commanding,
                          1st Bn. The Hampshire Regiment.

That operation order laid out the bare bones of the expedition, mincing no words nor minimising the difficulty of the task that lay ahead. Not truly a campaign in the sense that war-gamers know it, this operation can be completed in an evening. It provides an excellent evening or afternoon's project for a war-games club, with each member commanding a separate part of the British force, the tanks and armoured cars whilst others represent the leaders of assorted groups of tribesmen barring the way to their village.

The terrain is as shown on the map. As a practical note it may be mentioned that the various djebels or craggy heights were each mounted on separate irregular pieces of chipboard. Made from sacking soaked in polyfilla and allowed to dry in rough irregular shapes, each of the pieces contained innumerable nooks and crannies, crevices and valleys in which marksmen could be placed in small groups. The entire floor of the valley was covered with minute circles of paper taken from a desk device for punching holes in paper. A number of these paper circles had a cross marked on one side, placed face downwards; they were turned up as and when a man or vehicle touched them

—a cross denoted a land mine. It was unlikely that the weight of a man would explode one of these mine (it required a throw of 6 on a dice to do so) but the weight of a vehicle (a lorry, armoured car or tank) would certainly cause them to explode. As a vehicle proceeded over one of the 'stones' they were turned up as touched and when the cross was revealed a dice was thrown.

1 or 2—the mine failed to explode.
3 or 4—the vehicle disabled for one move.
5 or 6—the vehicle was permanently knocked out.

It is not the policy of this book to lay down the specific rules under which campaigns or battles are fought but there are some original or unusual factors involved in this operation which are considered to be of sufficient interest to be laid out in greater detail.

The heavy machine-gun in the armoured car fires by throwing six dice per move. These dice may be thrown together to represent 'blanketing' fire or one after the other as the gun traverses along a position. The effect of fire is represented by the following method:

|  | *Car moving* | | *Car stationary* | |
|  | *Massed* | | *Massed* | |
|  | target | *Deployed* | target | *Deployed* |
| 30 in. | $\frac{1}{2}$ score | $\frac{1}{4}$ score | $\frac{3}{4}$ score | $\frac{1}{2}$ score |
| 15 in. | $\frac{3}{4}$ score | $\frac{1}{2}$ score | Full score | $\frac{3}{4}$ score |

Half scores only count when target behind hard cover.

The tribesmen are adept in constructing protective cover from rocks and boulders—known as sangars, these small breast-works were made by pressing together small balls of brown plasticine to a sufficient height to allow men to be lying down or kneeling behind them for firing purposes.

The tanks carried 2 pdr. guns which could fire solid shot or high explosives. It was necessary at the beginning of each game-move for the commander of the tank to write down the type of projectile loaded into the gun. Rules must be made so that the effects of solid shot upon the mud walls of the fort or upon the stone sangars are realistically represented or the effects of H.E. bursting in the air over and above a position. The firing of the gun was represented by picking a point of aim and then throwing a dice. When a hit was scored a burst circle 3 in. in diameter was used and all covered were killed without saving throws. If men are under cover then the usual saving throws prevail but an air-bursting H.E. shell above a group of men sheltered behind a stone sangar would neutralise the effect of that cover.

|  | *Tank moving* | *Tank stationary* |
|---|---|---|
| 36 in. | Hit scored by 3. | Hit scored by 3 or 4. |
|  | 1 or 2=3 in. short. | 2 and 5 is 3 in. right or left. |
|  | 4 or 5=3 in. right or left. | 1 is 2 in. short. |
|  | 6=3 in. over. | 6 is 3 in. over. |
| 18 in. | Hit scored by 3 or 4. | Hit scored by 2, 3, 4, 5. |
|  | 2 and 5=3 in. right or left. | 1=3 in. short. |
|  | 1=3 in. short. | 6=3 in. over. |
|  | 6=3 in. over. | |

Some of the tribesmen carried petrol bombs—these men had a small plasticine ball fixed on to their rifle. If such a man was hit by rifle fire it was considered that the petrol bomb would either explode at the time or when he fell so that the man had no saving throw. A petrol bomb could be thrown from a range of 6 in. or dropped on vehicles. Its effects were simulated by throwing a dice:

1 or 2—Missed or failed to ignite.

3 or 4—Fire extinguished in 1 move without the vehicle being permitted to move.

5 or 6—The vehicle is totally destroyed and a dice must be thrown for each member of the crew to see whether he is saved or not.

At certain vantage points in the terrain boulders were placed that could be rolled down slops at vehicles or men coming up a path for example. These boulders were represented by balls of plasticine and were actually rolled down, two men were required to push them. If a vehicle was hit by a boulder a dice was thrown:

1 or 2—The vehicle was disabled for 2 moves.

3 or 4—The vehicle was disabled for 1 move.

5 or 6—The vehicle was undamaged.

In the event of a boulder running through a group of men, then every man knocked over by that boulder is considered to have become a casualty—a dice throw of 1, 2 or 3 means that he is killed, 4, 5 or 6 severely wounded sufficiently to require to be taken back to the First-Aid post.

The nature of the terrain meant that it was necessary for the tanks and the armoured car to move 'opened up' with the commander's head out of the turret. Naturally, this made the vehicle more vulnerable and tribesmen were allowed to fire upon it but 1 was deducted from each firing-dice to allow for the smallness of the target. In the event of a tank or armoured car commander being killed, a dice was thrown:

1 or 2—The tank or armoured car withdraws from the action.

3 or 4—The tank remains stationary for 1 move.

5 or 6—The tank or armoured car carries on in the normal manner.

It should have been mentioned earlier that all troops used in this operation were Airfix figures. The British troops consisted of those from the 8th Army box whilst the tribesmen were Bedouin Arabs. Every figure moved individually and each of them had a small piece of lead sheet glued to the underside of its base. Sheets of roofing lead can be purchased from any ironmongers and plastic Airfix figures so treated achieve a remarkable stability. In this particular operation involving rocky slopes and irregular terrain, it was possible for both British and tribesmen to stand on the roughest surfaces at what would normally be considered as impossible angles for model soldiers. The vehicles used were a Lesley Matchbox armoured car suitably converted to pre-World War II vintage and painted sand colour. The two tanks were 'mocked-up' vehicles made from parts of Airfix tank kits and vaguely resembled the Vickers Mk. IV tank.

No better way can be found of describing the manner in which this operation proceeded than by reproducing the report of the Officer Commanding the force as rendered on his return to Razmak.

*Subject : Punitive Expedition to destroy the Village and Fort of the Malik-Ghazi Tribe in the Pushna Valley.*

To: Officer Commanding 1st Bn. of The Hampshire Regiment.

From: Major J. B. Hollands, M.C., Officer Commanding Expedition.

The force under my command arrived at the south-eastern corner of the Pushna Valley at first light on the 5th November, 1936. In the far distance, north-west corner of the valley the

o

towers and walls of the mud fort could be seen rising above palm trees growing along the edge of the plateau upon which the fort was built. On examination, the Valley itself appeared to be more a series of small valleys or defiles rather than one large valley between rising hills. This fact is not clearly apparent from available maps and it is suggested that maps should be suitably amended for the benefit of possible future expeditions in this area.

At the entrance of the valley immediately in front of us as we entered rose Djebel Kournine, a rocky height with two peaks towering at either end and a sandy plateau that could only be entered on the East and West sides by two narrow, steeply rising tracks. A valley lay between the Kournine and Djebel Shatar which could only be approached via the Kournine or through two thick belts of palm trees and jungle that blocked either end. Jungle continued down to the South-West of the valley towards a winding and steep track that led up to the plateau on which the village and fort were situated. West of Djebel Shatar lay Djebel Omar running directly East and West and consisting of two almost unclimbable rocky ridges with a very narrow ravine stretching between them. South of Djebel Omar was Djebel Barga, consisting of a plateau surrounded on all four sides by unclimbable rock face with only a steep track as its sole entry on its Western face. The entire rear or Western end of the valley was completely closed by a rock face climbing high into the air to a large plateau which stretched far off out of sight. This plateau could only be reached by two very narrow winding rocky paths—one at the North and one at the South ends of the rock face. At the very Northern end of the plateau, behind a thick grove of palm trees and jungle, lay the mud fort and village of the Malik-Ghazi tribe.

Through the glasses it was possible to see tribesmen lying in position on the majority of the climbable rocky eminences. They were behind stone sangars and breastworks, positioned high above the narrow tracks through which the force had to pass. It was considered that the tribesmen in view were far from being the full strength of the tribe—a fact that was later confirmed when a large force of camel riders emerged from behind Djebel Barga and a similar-sized force of horsemen were hidden at the Western end of Djebel Omar. During the course of the operation three further bands of tribesmen, estimated to be about 25 to 30 men in each, were discovered congregating behind rock faces on the Djebel Shatar; in the valley between the ridges of Djebel Omar and in the jungle and palm trees to the South of the Omar.*

I commenced operations by ordering the tanks and armoured car to proceed along the Eastern border of the valley, closely followed by and shielding the remainder of the force in their trucks. On coming level with the rocky path leading up to the Djebel Kournine, the force came under fire from a party of tribesmen tucked away behind stone breastworks on the Kournine. Immediately I ordered Number 1 Platoon to debus and ascend the steep track to winkle the tribesmen out from their sandy plateau half way up the rocky eminence. I also ordered supporting fire from the tanks and the armoured car as Number 1 Platoon proceeded up the path.

As these vehicles passed the bottom of the rocky track leading up to the Kournine on their way North-West, a large boulder was rolled down the track to narrowly miss the leading vehicle.

The main party continued round the Northern end of the Kournine, still under fire from small parties of tribesmen on the eminence. At the same time Number 1 Platoon toiled upwards and was soon engaged in a sharp fire-fight and then a hand-to-hand combat with the tribesmen. Heavy casualties were sustained by this Platoon who were reduced to approximately half strength.

The advance continued slowly around the Northern end of the Kournine with infantrymen leading the force to probe for suspected mines—without success. Soon it became obvious that such a cautious method of approach was too time-consuming and it was agreed between myself and the Commanders of the armoured vehicles that they should proceed forward and that the mines should be considered as a calculated risk.

Once round to the West of Djebel Kournine, I had to decide whether to press down the valley between Djebel Omar and Djebel Barga or to move immediately South to the small valley between Djebel Shatar and Djebel Omar. I decided on the latter course which had the effect of marooning the tribesmen on Djebel Shatar so that they were unable to contact my force except by long-range fire. As soon as this fact was realised, the tribesmen on Djebel Shatar moved from their positions and

* These men were not placed in position but were represented by one man and a counter marked with the complete number of the party he represented.

climbed the winding track on the West side of Djebel Kournine where they engaged the remnants of Number 1 Platoon who had wiped out the force originally opposing them. Number 2 Platoon had been left at the Southern end of Djebel Kournine to cover the force in case of a flank or rear attack by tribesmen. I now consider that this expenditure of manpower, pinned down by a lesser number of tribesmen, was wasteful and unnecessary.

Moving down the valley between Omar and Shatar the force was dealt a severe blow when a tribesman dropped a petrol bomb on to the armoured car causing it to go up in flames. Almost at once, a light tank struck a mine and similarly brewed up. Showing the utmost disregard for the safety of himself and his vehicle, Sergeant Baker in command of the remaining tank pushed down the valley towards a patch of jungle South of Omar under a heavy fire from Shatar, Omar and the jungle itself.

Although the machine-guns and 2 pdr. anti-tanks and armoured car had been causing apparently severe casualties there seemed to be no diminishment of the enemy numbers.

At the Western end of Djebel Omar, a petrol bomb was dropped into a truck carrying Number 3 Platoon which went up in flames. Five men of the Platoon were killed. At this point Sergeant Baker in the light tank was shot through the head and killed. His replacement, Corporal Jennings, was similarly killed as soon as he assumed command and put his head through the open turret top. This left the tank with a one-man crew so that it was able to move or fire but not both at the same time.

At this stage it appeared likely that the force under my command would have to withdraw from the valley without achieving our objective. However, timely reinforcements consisting of two armoured cars of the 3rd Light Company came into the valley at our starting point. One of these cars came round the Northern end of the Kournine, following the route of the rest of the party and bringing with them the lorry-borne remainders of Numbers 1 and 2 Platoons. This force was sufficiently threatening to prevent the tribesmen on Djebel Barga or the camel reserves from moving out of their position to aid their comrades at the Southern side of the valley. The second armoured car skirted the jungle South-West of Kournine and, coming past the still burning vehicles in the valley between Shatar and Omar, similarly prevented the tribesmen on Shatar from moving down to attack our force.

Led by the light tank, the force reached the foot of the winding track at the South-Western corner of the valley and began to ascend. With the tank in front, the three lorries bearing the remainder of the infantry moved slowly forward whilst the two armoured cars remained at the foot of the track covering the party.

It seemed that the Arabs had expended all their petrol bombs so that there was no danger to fear from this quarter. At the same time a safe track through the mines was known so that there was little to prevent the force from reaching the plateau and destroying the fort and village. There appeared to be no tribesmen on the top plateau, either they had fled or had come down into the valley to defend their territory.

The fort and village were destroyed and the force returned without incident. Although there was still a considerable force of camel- and horse-mounted men in the valley they apparently did not fancy their chances of attacking us. There is no doubt that the arrival of the armoured cars under the able command of Sergeant Fremantle and Corporal Bates had made the difference between success and failure. Similarly, not realising that the tank was incapable of moving *and* firing, the tribesmen paid it the utmost respect.

The expedition reached Razmak on its return journey at 2200 hours on the 5th November, 1936.

*Casualties*

A Company—*Killed:* 1 Officer, 1 Sergeant, 2 Corporals and 12 men. *Wounded:* 1 Officer, 1 Corporal and 17 riflemen.

3rd Light Company Royal Tank Regiment—*Killed:* 1 Lieutenant, 2 Sergeants, 1 Corporal and 2 Troopers. *Wounded:* 2 Troopers.

Razmak Cantonment.

6 Nov. 1936.

J. B. HOLLANDS
Major Commanding A Company,
1st Bn. The Hampshire Regiment.

# 28

# A Landing in Force
## (Coast of France 1944)

Sometimes campaigns are not born—they just grow! Perhaps by accident or intention, a war game takes place in which a convoy is attacked, a frontier violated or a treaty broken, so providing an excuse for the indignant injured party to summon up their forces and attempt to achieve revenge. Perhaps no more real-life logical opening to a campaign occurred than on the 6th June, 1944, when the Allied forces landed on the enemy-held coast of Normandy. Similarly, for an invasion force to attempt to land on a table-top terrain laid out to represent a coastline is to stage a battle that, if successful, will be the first shots fired in a campaign.

Veteran modernist war-gamer Ron Miles of Southampton carried out such a battle and, at the time of writing, it is not known whether a full-scale operation followed or if the invaders were repulsed or sufficiently discouraged to await another day.

The operations centred round the mythical 1st British Combat Group, formed from all arms and normally independent although subservient to normal commands. Commanded by a lieutenant-general, the force consisted of:

2 Battalions of infantry, one of which was motorised.
1 Tank regiment.
1 Armoured car squadron.
1 Self-propelled anti-tank gun battery.
1 Light anti-aircraft battery.
1 Field regiment of artillery.
   Engineers and armoured special units, i.e. Flail, Ark, bridging and recovery units.

A squadron of fighters for close support work was attached and a destroyer detailed, after initial bombardment for close fire support. An additional glider-borne company and airborne paratroops made up the complete unit.

The infantry battalion was composed of four rifle companies each of 26 riflemen, 2 L.M.G. teams of two, an officer and W/OP. In addition, there were an H.Q. support company, 4

M.M.G.s, 4 mortars and OP, 2 6 pdr. anti-tank guns equipped with half tracks and universal carriers. The tank regiment was a composite unit formed of two squadrons of three Shermans each, one squadron of Fireflys, one squadron of Churchills. Of the specials—one Churchill was a Crocodile flame-thrower

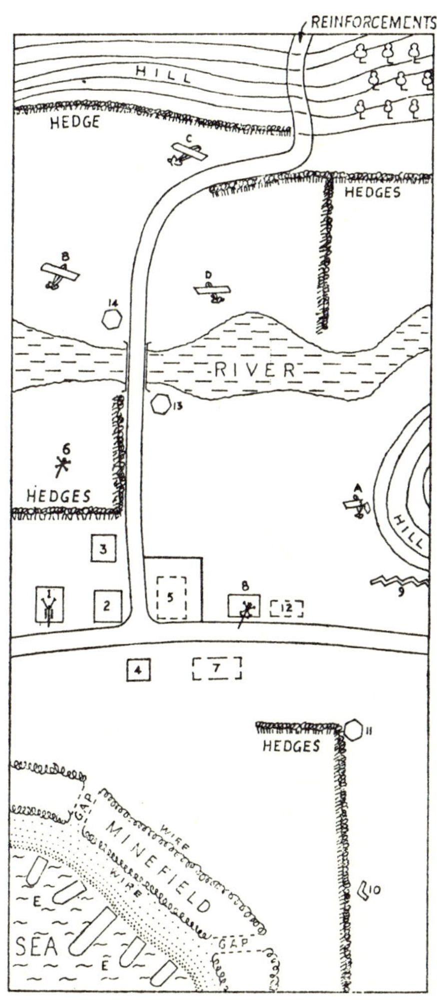

and the bridge carrier was equipped with a petard mortar. Two M 10s made up the SP anti-tank battery.

This unit was given the task of seizing the bridge over the river on the extreme left of the landing area about a mile from the beach. They were to follow up quickly to take the high ground just beyond the river, so facilitating the build up for a later breakout.

According to intelligence, the strength of the defences was considered to be fairly strong although under-manned. A reinforcement of both armour, infantry and possible nebel-werfers could be expected. Artillery and Luftwaffe reaction could be swiftly anticipated. A minefield had been laid and work was in progress on Dragons' Teeth. Pillboxes and strong-points had been built and the only possible approach was by landing in a small bay, passing through the village along the road to the bridge.

The battle was fought on a table 10 ft. by 4 ft. 6 in., with terrain made from polystyrene tiles built up and painted into contours, Airfix buildings, a home-made bridge and pillboxes, rubberised horse hair dyed for hedges and trees, aircraft mounted on thin wire strands and artillery was 'off the table' (for explanation of this method see page 51 of the book *Advanced War Games*).

Aircraft were allowed a time limit of four moves over the table, moving at the maximum rate of 60 in. per game-move. They could not fire during the whole period whilst over the table, being permitted to fire for one complete move only or in bursts of four phases, into which every full move was divided. (For full details of Ron Miles' method of simulating aircraft in war games see chapter 6 of the book *Air War Games*).

Infantry moved at $4\frac{1}{2}$ in. per move; tanks and vehicles moved at suitable rates, i.e. heavy tanks 6 in. per move and half-tracks on a road could manage 18 in. Armament ranges varied from 10 in. for a rifle whilst tank guns fired, according to calibre, etc., up to 40 in. Tanks carried only a limited number of rounds, being split into H.E., anti-tank or smoke at the commander's discretion.

During the forces' approach to the landing place, a dice was thrown to discover if any enemy forces were in the vicinity. It turned out to be only a light unit who panicked and scampered away without informing their headquarters of the approaching armada. This meant that the first the Germans knew of the invasion was when the initial bombardment hit them. For quickness and ease this is done by dice, the areas being divided into sections and the strongpoints numbered from 1 to 6. After deciding whether or not the bombardment was successful, a dice was tossed to discover the amount of damage done to the defences. It so worked out that the minefield was gapped in two places, part of the wire was also gapped and two of the strongpoints were knocked out.

As the landing craft hit the beach, the glider-borne troops swept in overhead. With so much to engage, the German forces were caught in two minds and exposed their positions. The 88 mm gun took on an anti-aircraft role but missed with its first shot, as did all the other guns. Second shots were more successful, hitting a glider and damaging it. The Germans were now heartened by the arrival of an M.E. 109 which promptly shot down the glider before tangling with a Spitfire so that each damaged the other in passing. The Spitfire went in low to strafe German forces by the bridge but did no damage. The M.E. 109 strafed the British troops disembarking from the landing craft, killing one man before being shot down by a disembarked Bofors gun. The survivors of the shot-down glider were taken prisoners by the Germans. The disembarkation proceeded smoothly with both tanks and infantry finding the narrow gaps in the minefields. A Flail tank started to work on the minefields aided by some engineers. Suddenly a German artillery stonk fell in the area, damaging landing craft and causing casualties in addition to destroying a carrier and an armoured car. These artillery stonks were handled in the following manner:

Multiply the number of firing guns by 4 (the number of rounds fired in a minute) so that 6 guns multiplied by 4 equal 24 rounds in the stonk. Then throw a dice for percentage (the figures on the dice being previously nominated to represent varying percentages from 10 to 100). Say 75% is attained— 75% of 24 is 18 rounds on the target. Next it remains to decide where the shells land which can be done by a burst circle or artillery-spider device (described in the London War Games Section Rules of 'Bish' Iwaszko).

The build-up on the beach was proceeding but the breakout was strangely lethargic, being hampered by accurate but ineffective fire from the 88 mm and 75 mm guns, the minefield and wire. While the glider-borne troops touched down and debussed, it was here that the objective was to be won or lost. Unfortunately, this was also the place in which things did not go right for the British. First the troops came under heavy small-arms fire from the pillboxes which caused several casualties. Worse still, the first of the reinforcements—a German tank company—arrived on the top of the hill and opened fire on the 6 pdr. anti-tank gun. They missed the gun but hit the towing vehicle and knocked it out. Desperately, the gunners tried to deploy their gun but the concentrated fire of the tanks was too much for it and the one remaining gunner was forced to retreat and leave his weapon. The Panzers closed in on the sky troops just as the sea-borne forces started their breakout. The first Sherman tank emerged from the gap and was knocked out. The British forces laid a smokescreen and a Churchill tank rushed the minefield but lost a track. Further round the bay some infantry got through, followed by a Churchill Crocodile, while an armoured car squeezed through the gap near the knocked-out Sherman. Finally, the Flail tank cleared a path through the minefield but inexplicably pushed on instead of widening the gap. More infantry got through and came under small-arms fire. As the invading forces applied pressure so the advanced German posts gradually fell back.

The Germans' 75 mm gun was knocked out, allowing the British troops to get on to the road leading to the bridge and so outflanking the 88 mm gun. Unfortunately it was too late for at that moment, having finally overcome the airborne troops, the first German tank arrived on top of the bridge and sat astride the road completely blocking it.

Well, that is as far as Ron Miles' account goes—it is left to the reader to decide whether or not the British commander would be justified in carrying on the invasion or calling it off and trying another day.

# 29

# Guards *v.* Panzer Grenadiers
### (France 1944)

A map as shown was drawn to the scale of 1 in.＝12 in. on the war-games table with 6 in. on the map＝1 mile. The complete map was formed of 3 by 3 war-games tables each measuring 6 ft. by 6 ft. Movement is subject to the following provisos:

Woods can only be entered by infantry.

All buildings can be entered and will give shelter.

Hills can be climbed at half speed; a railway embankment is climbable at the same rate.

The river can only be crossed by the bridges but may be traversed by assault boats, pontoon bridges, bridge-laying tanks (Arks) or other engineering means. (It will be necessary to specify the time taken to bridge the river or to cross it by other means. Similarly, allocations of bridging equipment must be specified for one or both forces before the commencement of the battle.)

*Narrative*

It is September 1944 and the Hermann Goering Panzer Grenadiers, accompanied by armour with artillery, have made a sudden breakthrough on a small area of the front and in an attempt to reach the junction town of Louisville. To cope with this situation, the British Command has pushed forward the 2nd Battalion Coldstream Guards, together with a squadron (9 tanks) of the 3rd Royal Tank Regiment and some artillery. The forces used in this action were those mentioned in chapter 15 of the book *Advanced War Games*. This section detailed a method of forming from Airfix figures a battalion of infantry split up into its respective companies, platoons and sections. In chapter 28 of the book *Advanced War Games* there is described a demonstration battle between the same forces that are taking part in the action now being described.

At the commencement of the battle the British can lay down their entire force *anywhere* they desire in squares D, E, F, G, H and I. The German forces, split up as desired, may come in anywhere on the northern baseline of the map. At the start, with the British in the squares nominated above, the Germans carry out the first of their map-moves. Simultaneously the British move on their map, probably sending patrols into squares A, B and C to discover German intentions.

This campaign can be fought by two war-gamers with a third man acting as umpire. On the other hand, it is ideally suited as a Club Project with literally any number of people participating, taking command of patrols and individual groups, etc. In the case of two war-gamers with an umpire, they make their map-

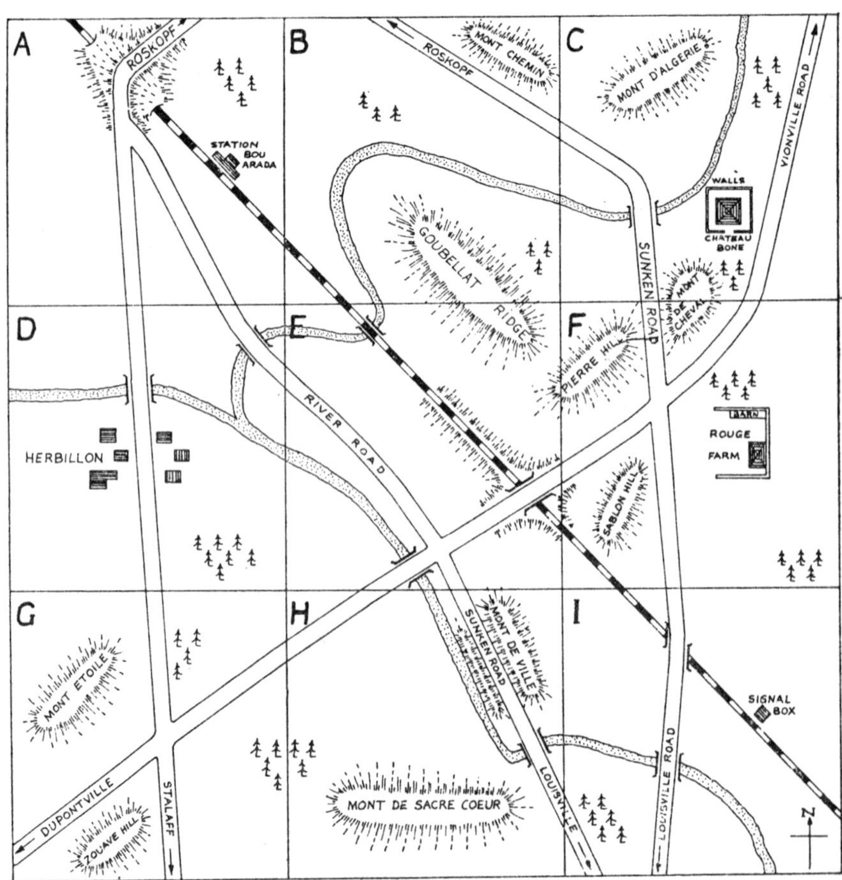

moves with chinagraph pencil on a transparent plastic map cover and then hand these covers to the umpire who places them upon his master map and decides whether or not a contact has been made or decides any other situation that may have arisen.

The two principal features of this campaign which give it originality and unusual interest lay in the use of patrols and in the fact that these patrols have to report back to their main bodies by runner or radio. Indeed, this campaign is what might almost be termed as an exercise in communications because movement is only carried out and orders are only amended on receipt of information by radio or runner.

The radio sets used in this game are as follows:

88 set (range 1 mile) carried by an infantry platoon or company and by an artillery battery.

31 set (with a range of 3 miles) carried by armoured fighting vehicles (tanks, armoured cars and recce vehicles), by an infantry battalion command vehicle, by an infantry commander and by an artillery battery.

19B set (with a range of 1 mile) carried by AFVs and by infantry battalion command vehicle.

The overall commander in the area will be able to receive messages from all arms irrespective of the radio set carried.

Each piece of radio equipment must be tested at the beginning of each move to see if it is 'on net' with other sets. A dice throw of 3, 4, 5 and 6 means that it is in contact, whilst 1 or 2 means that the set is out of contact for that move. When the radio operator working a set is killed or wounded, his equipment must be diced for—5 or 6 means that the equipment is still workable; 1, 2, 3 or 4 means that the equipment is out of action.

Orders will be written in the normal way and if, for any reason, a unit or a group is required to deviate from these written orders, supplementary orders must be given to it either by runner who moves at double infantry rate and who must always be tested each move—a dice throw of 3, 4, 5 or 6 indicates that the runner gets through whilst 1 or 2 means that the runner gets lost; or by radio when a throw of 2, 3, 4, 5 and 6 on the dice means that the message gets through while 1 means that the set is not working.

When a patrol or any force or vehicle comes within the ranges of vision detailed below (when working on the map in the initial stages of the campaign) then information may be sent back by runner or radio. In these two cases a dice throw of 1 or 2 means that the runner gets lost whilst a dice throw of 1 means that the radio is out of action. Any group containing a radio operator that is within sight of the enemy and is fired upon so that the radio operator is killed must decide by means of dice or some similar method whether or not they have got their message off before the casualty occurred.

*Range of vision* (subject to there being no obstacles in between)

| A patrol can: | Daybreak or dusk | Daylight |
|---|---|---|
| Distinguish vehicles en masse | $\frac{1}{2}$ mile | 1 mile |
| Distinguish vehicles individually | $\frac{1}{3}$ mile | $\frac{3}{4}$ mile |
| Distinguish troops en masse | $\frac{1}{3}$ mile | $\frac{3}{4}$ mile |
| Distinguish troops individually | $\frac{1}{4}$ mile | $\frac{1}{2}$ mile |
| Distinguish concealed troops | | $\frac{1}{4}$ mile |
| Distinguish concealed vehicles or guns | 220 yards | $\frac{1}{3}$ mile |

*Ranges at which patrols can be sighted*

| | Daybreak or dusk | Daylight |
|---|---|---|
| Infantry patrol in the open | | $\frac{1}{4}$ mile |
| Infantry patrol under cover | | 220 yards |
| Armoured Car or recce vehicle in open | $\frac{1}{4}$ mile | $\frac{1}{2}$ mile |
| Armoured Car or recce vehicle under cover | 220 yards | $\frac{1}{4}$ mile |
| Tank in open | $\frac{1}{2}$ mile | $\frac{3}{4}$ mile |
| Tank, hull-down | 220 yards | $\frac{1}{4}$ mile |

Weather conditions will affect these distances—on a rainy or dull day all distances are down by 25%.
On a misty day the range of vision will be the same as at dusk.
On a foggy day all distances are cut down by 75%.
In snowy conditions add 50% to all distances.

For the purposes of indicating these distances on the map

6 in.$=$1 mile.
$4\frac{1}{2}$ in.$=\frac{3}{4}$ mile.
3 in.$=\frac{1}{2}$ mile.
2 in.$=\frac{1}{3}$ mile.
$1\frac{1}{2}$ in.$=\frac{1}{4}$ mile.
$\frac{3}{4}$ in.$=$220 yards.

An interesting measure of realism can be brought into this game when it is being fought as a Club Project. Then an individual will be in command of a patrol and will make his own forward map-movements, being responsible for sending back the necessary messages to another war-gamer who is in command of the body in the rear from which the patrol has gone. Until the message is received back at normal rates of movement and subject to the conditions laid down above, then the commander of the main force has no information as to the findings of his patrol. A patrol can indicate the exact location of an enemy force by means of giving its position by a Map Reference. All Ordnance maps are gridded with numbered squares by which it is possible to give a vertical and horizontal numbered reference that will exactly pin-point a spot on the map. The home-drawn map we are using is not similarly gridded if only for the simple reason that such a grid would take a long time to draw and would also greatly obscure the map. An ideal way of reproducing a Map Reference system with a home-made map is to grid the transparent plastic cover upon which map-movements are drawn with a wax pencil. In this case it will be necessary to grid the cover in 1 in. squares numbered vertically from top to bottom 0 to 18 and horizontally from 0 to 18 also. To demonstrate the method of working, take a look at the diagrammatic map accompanying this chapter and find the bridge that crosses the railway line on the embankment in square E. Any force on that bridge would be on point 1011 because the bridge is on the tenth line from the top of the map and the eleventh line from the map's western border.

Thus a patrol sighting an enemy force in this position would report back by radio:

'Enemy tanks accompanied by lorried infantry crossing railway bridge on embankment at point 1011.'

Immediately on receipt of this message, the commander of the parent body would place his transparent grid cover on his map, run his finger swiftly across line ten and down line eleven to arrive at the bridge over the embankment which would immediately pin-point the enemy and inform him of their whereabouts.

The amount of artillery given to each side is a matter of choice, bearing in mind that the range of World War II artillery was considerable so that guns were not forward with the infantry (with the possible exception of anti-tank guns). In this particular campaign, at the commencement British artillery were on call anywhere in squares G, H and I. Their fire could be called down by a message giving them a Map Reference. The means of simulating the effect of that fire is fully described on pages 49, 50 and 51 of the book *Advanced War Games*. Included in this description are methods of using fire plans, off-table shoots, the use of telephone lines from batteries to observers, blanketing an area with fire, etc., etc. The German artillery are initially off the table, immediately North of the top of the map until action moves into squares D, E and F, when the German artillery may move forward into squares A, B and C.

Purposely, many of the finer points and details of modern war-gaming have been omitted in this campaign. The purpose of this description is to indicate lines along which it can be fought and to stimulate war-gamers to approach their war-gaming from what may optimistically be claimed as a more realistic approach. It is obvious that the British force will have to be disposed in the middle three squares of the map, ready to move into action against any German thrust after determining whether it is the main effort or a feint. Once a contact has been made then the necessary table will be laid and, if desired, either commander may bring in reinforcements from other squares who will move at the scaled rates of movement on the map until such time as they cross the border line and come into the square which is actually represented on the war-games table when they will take an active part in the battle.

As a means of concluding this campaign, it may be revealed that immediately South of the bottom border of the map is an open plain leading to the junction town of Louisville. The British have no chance whatsoever of holding up the enemy on that flat and exposed stretch of land. This means that once the British have been outflanked or pushed off the map then they must retire back behind Louisville and they will be considered to have lost the campaign.